Monographs in Oral Science

Vol. 24

Series Editors

A. Lussi   Bern

M.C.D.N.J.M. Huysmans   Nijmegen

H.-P. Weber   Boston, Mass.

Saliva: Secretion and Functions

Volume Editors

Antoon J.M. Ligtenberg Amsterdam
Enno C.I. Veerman Amsterdam

56 figures, 35 in color, and 15 tables, 2014

Basel · Freiburg · Paris · London · New York · Chennai · New Delhi ·
Bangkok · Beijing · Shanghai · Tokyo · Kuala Lumpur · Singapore · Sydney

Antoon J.M. Ligtenberg
Academic Centre for Dentistry Amsterdam (ACTA)
Gustav Mahlerlaan 3004
NL–1081 LA Amsterdam
The Netherlands

Enno C.I. Veerman
Academic Centre for Dentistry Amsterdam (ACTA)
Gustav Mahlerlaan 3004
NL–1081 LA Amsterdam
The Netherlands

Library of Congress Cataloging-in-Publication Data

Saliva : secretion and functions / volume editors, Antoon J.M. Ligtenberg,
Enno C.I. Veerman.
     p. ; cm. -- (Monographs in oral science, ISSN 0077-0892 ; vol. 24)
  Includes bibliographical references and indexes.
  ISBN 978-3-318-02595-8 (hard cover : alk. paper) -- ISBN 978-3-318-02596-5
(e-ISBN)
  I. Ligtenberg, Antoon J. M., editor. II. Veerman, Enno C. I., editor. III.
Series: Monographs in oral science ; v. 24. 0077-0892
  [DNLM: 1. Saliva. 2. Salivary Glands. W1 MO568E v. 24 2014 / QY 125]
  QP191
  612.3'13--dc23
                                                    2014009660

Bibliographic Indices. This publication is listed in bibliographic services, including Current Contents® and Index Medicus.

Contents

List of Contributors

Thomas Arnebrant
Department of Biomedical Sciences
Faculty of Health and Society
Malmö University
SE–205 06 Malmö (Sweden)
E-Mail thomas.arnebrant@mah.se

Watcharapong Aroonsang
Department of Prosthetic Dentistry
Faculty of Odontology
Malmö University
SE–205 06 Malmö (Sweden)
E-Mail awatcharapong@gmail.com

Hendrika Bootsma
Department of Rheumatology and Clinical
Immunology
PO Box 30001
NL–9700 RB Groningen (The Netherlands)
E-Mail h.bootsma@umcg.nl

Jos A. Bosch
Department of Clinical Psychology
University of Amsterdam
NL–1018 XA Amsterdam (The Netherlands)
E-Mail j.a.bosch@uva.nl

Henk S. Brand
Department of Periodontology and Oral Biochemistry
Academic Centre for Dentistry Amsterdam (ACTA)
Gustav Mahlerlaan 3004
NL–1081 LA Amsterdam (The Netherlands)
E-Mail H.Brand@acta.nl

Paulo Camargo
UCLA School of Dentistry
10833 Le Conte Avenue, CHS 53-038
Los Angeles, CA 90095 (USA)
E-Mail pcamargo@dentistry.ucla.edu

Guy H. Carpenter
Salivary Research Unit
Floor 17 Tower Wing
King's College London Dental Institute
Guy's and St Thomas' Hospitals
Great Maze Pond, London SE1 9RT (UK)
E-Mail guy.carpenter@kcl.ac.uk

Konstantina Delli
Department of Oral and Maxillofacial Surgery
PO Box 30001
NL–9700 RB Groningen (The Netherlands)
E-Mail k.delli@umcg.nl

Matthew P. Hoffman
NIDCR, NIH, Bldg 30/Rm 433
30 Convent Drive, MSC 4370
Bethesda, MD 20892-4370 (USA)
E-Mail mhoffman@mail.nih.gov

Kyle V. Holmberg
NIDCR, NIH, Bldg 30/Rm 429
30 Convent Drive, MSC 4370
Bethesda, MD 20892-4370 (USA)
E-Mail drkyle.holmberg@gmail.com

Frans G.M. Kroese
Department of Rheumatology and Clinical
Immunology
PO Box 30001
NL–9700 RB Groningen (The Netherlands)
E-Mail f.g.m.kroese@umcg.nl

Antoon J.M. Ligtenberg
Department of Periodontology and Oral Biochemistry
Academic Centre for Dentistry Amsterdam (ACTA)
Gustav Mahlerlaan 3004
NL–1081 LA Amsterdam (The Netherlands)
E-Mail A.Ligtenberg@acta.nl

Patricia Lima
UCLA Oral/Head and Neck Oncology Research Center
UCLA School of Dentistry
10833 Le Conte Avenue, CHS 73-034
Los Angeles, CA 90095 (USA)
E-Mail patyolima_1@hotmail.com

Liselott Lindh
Department of Prosthetic Dentistry
Faculty of Odontology
Malmö University
SE–20506 Malmö (Sweden)
E-Mail liselott.lindh@mah.se

Eric Neyraud
INRA, Centre des Sciences du Goût et de l'Alimentation
17, rue Sully
FR–21065 Dijon Cedex (France)
E-Mail eric.neyraud@dijon.inra.fr

Arie V. Nieuw Amerongen
Department of Periodontology and Oral Biochemistry
Academic Centre of Dentistry Amsterdam (ACTA)
Gustav Mahlerlaan 3004
NL–1081 LA Amsterdam (The Netherlands)
E-Mail A.vannieuwamerongen@kliksafe.nl

Gordon B. Proctor
Salivary Research Unit, Floor 17 Tower Wing
King's College London Dental Institute
Guy's and St Thomas' Hospitals
Great Maze Pond, London SE1 9RT (UK)
E-Mail gordon.proctor@kcl.ac.uk

Christopher A. Schafer
UCLA Oral/Head and Neck Oncology Research Center
UCLA School of Dentistry
10833 Le Conte Avenue, CHS 73-034
Los Angeles, CA 90095 (USA)
E-Mail caschafer1@ucla.edu

Jason J. Schafer
Department of Pharmacy Practice
Jefferson School of Pharmacy
Thomas Jefferson University
Philadelphia, PA 19107 (USA)
E-Mail Jason.Schafer@jefferson.edu

Francisco J. Silvestre-Donat
Special Care Patients
Clinica odontológica universitaria
Gascó Oliag 1
ES–46010 Valencia (Spain)
E-Mail francisco.silvestre@uv.es

Javier Silvestre-Rangil
Special Care Patients
Clínica odontológica universitaria
Gascó Oliag 1
ES–46010 Valencia (Spain)
E-Mail javier.silvestre@uv.es

Javier Sotres
Department of Biomedical Sciences
Faculty of Health and Society
Malmö University
SE–205 06 Malmö (Sweden)
E-Mail javier.sotres@mah.se

Fred K.L. Spijkervet
Department of Oral and Maxillofacial Surgery
PO Box 30001
NL–9700 RB Groningen (The Netherlands)
E-Mail f.k.l.spijkervet@umcg.nl

Wim van 't Hof
Department of Periodontology and Oral Biochemistry
Academic Centre for Dentistry Amsterdam (ACTA)
Gustav Mahlerlaan 3004
NL–1081 LA Amsterdam (The Netherlands)
E-Mail wimvanthof@hetnet.nl

Enno C.I. Veerman
Department of Periodontology and Oral Biochemistry
Academic Centre for Dentistry Amsterdam (ACTA)
Room 12-N37
Gustav Mahlerlaan 3004
NL–1081 LA Amsterdam (The Netherlands)
E-Mail e.veerman@acta.nl

Arjan Vissink
Department of Oral and Maxillofacial Surgery
PO Box 30001
NL–9700 RB Groningen (The Netherlands)
E-Mail a.vissink@umcg.nl

David T.W. Wong
UCLA Oral/Head and Neck Oncology Research Center
UCLA School of Dentistry
10833 Le Conte Avenue, CHS 73-034
Los Angeles, CA 90095 (USA)
E-Mail dtww@ucla.edu

Maha Yakob
UCLA Oral/Head and Neck Oncology Research Center
UCLA School of Dentistry
10833 Le Conte Avenue, CHS 73-034
Los Angeles, CA 90095 (USA)
E-Mail myakob@ucla.edu

Gleb E. Yakubov
School of Chemical Engineering
University of Queensland
St. Lucia, QLD 4072 (Australia)
E-Mail g.yakubov@uq.edu.au

Preface

We are glad to present this monograph on saliva. It provides a comprehensive overview of the latest developments in salivary research by some of the world's leading experts in the field. One of the most fascinating features of saliva and the oral cavity is that almost all aspects of life science research are applied to the oral cavity. However, the importance of saliva goes further than new technical developments in science. Health professionals are more and more aware of the importance of saliva for oral health and well-being. As saliva secretion is steadily compromised with advancing age, this becomes a factor of concern in societies with an aging population, especially with a growing number of people who keep their own teeth. In addition, the numerous functions of saliva, like antimicrobial activity, lubrication, wound healing and its role in taste experience, are only truly recognized when saliva secretion is hampered. Moreover, as a new line of research, salivary glands might even be regenerated with modern techniques like stem cell therapy.

In medical diagnostics, saliva shows its value as a safe and economical alternative to blood. Road side tests for alcohol and drugs as well as various home tests are already applied on a large scale in various countries. Also more advanced diagnostics, like the diagnosis of tumors, is possible by using saliva, because RNA and DNA are well preserved in saliva.

In the last sections of this monograph, saliva-related disorders are discussed. These may have to do with too much saliva, which results in drooling, or too little saliva, which may result in xerostomia. Also, complaints like infections of the salivary glands, sialoliths and mucoceles are discussed.

This publication is not only recommended to basic scientists working in the field of oral biology, but also to dental students, dentists and health professionals who want to know more about one of the most underestimated bodily fluids.

Antoon J.M. Ligtenberg, Amsterdam
Enno C.I. Veerman, Amsterdam

Ligtenberg AJM, Veerman ECI (eds): Saliva: Secretion and Functions.
Monogr Oral Sci. Basel, Karger, 2014, vol 24, pp 1–13 (DOI: 10.1159/000358776)

Anatomy, Biogenesis and Regeneration of Salivary Glands

Kyle V. Holmberg · Matthew P. Hoffman

Matrix and Morphogenesis Section, Laboratory of Cell and Developmental Biology, National Institute of
Dental and Craniofacial Research, National Institutes of Health, Bethesda, Md., USA

Abstract

An overview of the anatomy and biogenesis of salivary
glands is important in order to understand the physiology, functions and disorders associated with saliva. A major disorder of salivary glands is salivary hypofunction
and resulting xerostomia, or dry mouth, which affects
hundreds of thousands of patients each year who suffer
from salivary gland diseases or undergo head and neck
cancer treatment. There is currently no curative therapy
for these patients. To improve these patients' quality of
life, new therapies are being developed based on findings in salivary gland cell and developmental biology.
Here we discuss the anatomy and biogenesis of the major
human salivary glands and the rodent submandibular
gland, which has been used extensively as a research
model. We also include a review of recent research on the
identification and function of stem cells in salivary glands,
and the emerging field of research suggesting that
nerves play an instructive role during development and
may be essential for adult gland repair and regeneration.
Understanding the molecular mechanisms involved in
gland biogenesis provides a template for regenerating,
repairing or reengineering diseased or damaged adult
human salivary glands. We provide an overview of 3 general approaches currently being developed to regenerate damaged salivary tissue, including gene therapy,
stem cell-based therapy and tissue engineering. In the
future, it may be that a combination of all three will be
used to repair, regenerate and reengineer functional salivary glands in patients to increase the secretion of their
saliva, the focus of this monograph.

Salivary Gland Anatomy

The three pairs of major salivary glands in humans are the parotid (PG), submandibular (SMG)
and sublingual glands (SLG). The anatomical architecture of all three glands is essentially the
same: an arborized ductal structure that opens
into the oral cavity with secretory end pieces, the
acini, producing saliva. The acinar cells are surrounded by an extracellular matrix, myoepithelial
cells, myofibroblasts, immune cells, endothelial
cells, stromal cells and nerve fibers. The ducts
transport and modify the saliva before it is excret-
ed into the oral cavity through the excretory duct.
Stensen's duct is the main excretory duct of the
PG and enters the oral cavity in the buccal mu-

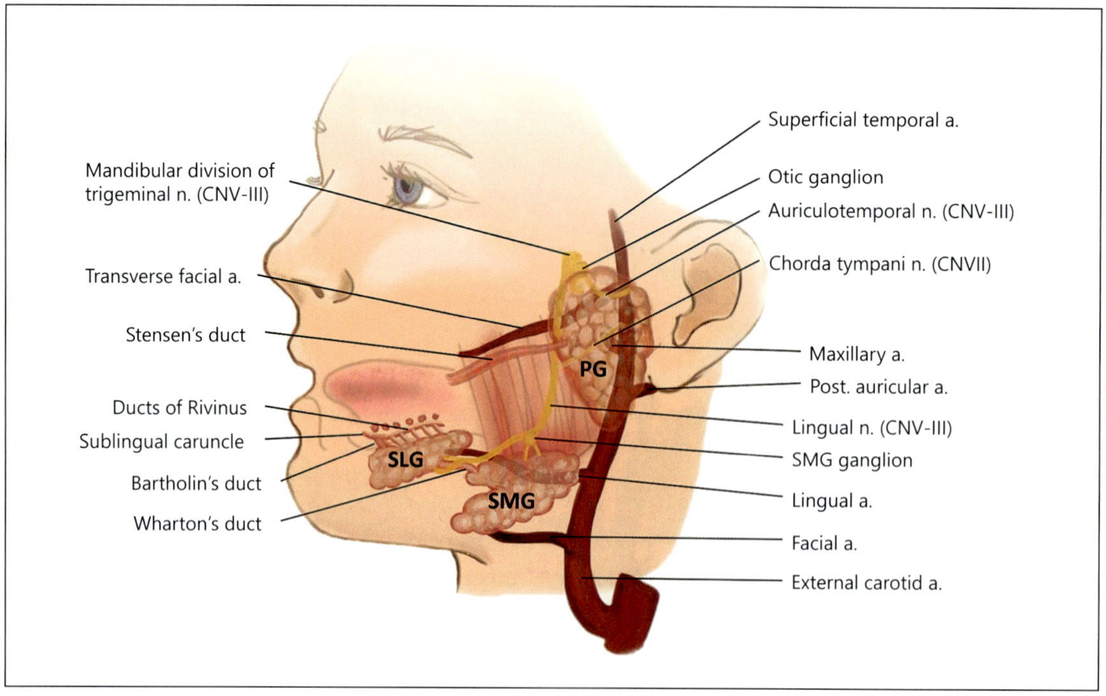

Fig. 1. Overview of salivary gland anatomy. The three major salivary glands are the PG, SMG and SLG. Stensen's and Wharton's ducts are the main excretory ducts of the PG and SMG, respectively. Blood supply is mainly provided to the PG by the transverse facial artery and to the SMG by the facial artery. Parasympathetic innervation arises from post-ganglionic nerves from the otic and submandibular ganglia. The otic ganglion is associated closely with the mandibular division of the trigeminal nerve (CNV-III), and the submandibular ganglion is next to the lingual nerve. Sympathetic postganglionic nerves (not shown) arise from the superior cervical ganglion and innervate the glands along blood vessels.

cosa near the second maxillary molar after crossing the masseter muscle and penetrating through the buccinator muscle. Wharton's duct is the main excretory duct of the SMG, which opens into the oral cavity under the tongue by the lingual frenum at a structure called the sublingual caruncula. The SLG has small ducts called ducts of Rivinus and a common duct, Bartholin's duct, which connects with Wharton's duct at the sublingual caruncula (fig. 1).

The major salivary glands are highly vascularized and innervated. The transverse facial artery emerges from the superficial temporal artery to provide blood supply to the PG and traverses along Stensen's duct. The facial artery, a branch of the external carotid artery, brings blood supply to

the SMG and passes through the gland capsule before crossing the inferior border of the mandible. The facial nerve (cranial nerve, CN, VII) is closely associated with the PG capsule, which also contains lymph nodes and is continuous with the superficial layer of deep cervical fascia. Facial nerve injury and resulting hemifacial paralysis constitute a significant risk of surgeries for PG tumor resection. The lingual nerve is closely associated with Wharton's duct in the floor of the mouth. Therefore, lingual nerve injury is a possible complication of surgical exploration of the floor of the mouth for removal of salivary stones. The capsule of the SMG is part of the superficial layer of deep cervical fascia. Lymph nodes are not within the capsule of the gland but are adjacent in

Holmberg · Hoffman

the submandibular triangle, an anatomic region formed by the boundaries of the inferior border of the mandible and anterior and posterior bellies of the digastric muscle [1, 2].

Saliva has multiple functions that include lubrication of the oral cavity to enable talking, swallowing, eating, tasting, dental health and maintaining oral homeostasis, while also providing protective functions and aiding in digestion. Many of these important functions will be covered in the papers by Lindh et al., van 't Hof et al., Brand et al., Neyraud and Yakubov [this vol., pp. 30–39, 40–51, 52–60, 61–70, 71–87] of this monograph. The different types of acinar cells in each gland result in different types of saliva. The PG is composed of serous acini and produces watery serous saliva. The SMG and SLG are both mixed glands containing both mucous and serous acini. The SMG has a majority of serous acinar cells with fewer mucous cells, whereas the SLGs are composed of a majority of mucous acinar cells. The major salivary glands in a healthy adult produce over 90% of saliva. In addition, there are minor glands, which are found in the palate and are widely distributed across the oral mucosa [3]. The secretion of saliva is stimulated by both the parasympathetic and sympathetic branches of the autonomic nervous system, and will be covered in the paper by Proctor and Carpenter [this vol. pp. 14–29] in this monograph.

The anatomy of the autonomic innervation of the major salivary glands is important to understand autonomic effects on not only salivation, but also biogenesis [4]. Briefly, parasympathetic stimulation results in serous, or watery, salivary secretion and ion secretion, whereas sympathetic stimulation increases the secretion of proteins. The cell bodies of the parasympathetic nerves that stimulate the PG are near the gland in the otic ganglion. The PG is innervated by postganglionic fibers that join the auriculotemporal nerve of the cranial nerve (CNV-III) [5, 6]. The cell bodies of the parasympathetic nerves that stimulate the SMG and SLG are located in the submandibular ganglia,

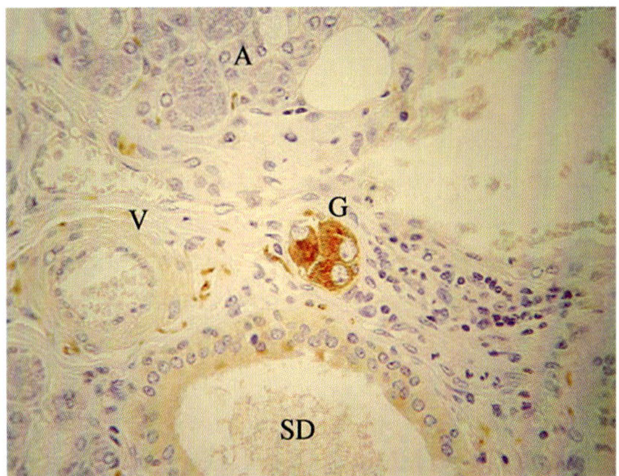

Fig. 2. The human SMG contains small ganglia within the gland stroma. Three synaptophysin-positive nerve cell bodies are arranged in a small ganglion (G) near acini (A), a striated duct (SD) and a blood vessel (V). Figure from figure 3 in Tosios et al. [9].

which are within the gland. The SMG and SLG are innervated by postganglionic fibers that stimulate saliva secretion and innervate myoepithelial cells [7]. The submandibular ganglia's preganglionic parasympathetic fibers are carried by the chorda tympani of CNVII, which joins the lingual nerve of CNV-III in the infratemporal fossa and then synapses at the submandibular ganglia [5, 8]. Importantly, there are also nerve cell bodies and small ganglia within the stroma of the SMG [9], although the functional significance of these ganglia is not clear (fig. 2). Interestingly, when SMG autotransplantation is used to treat absolute tear deficiency, the transplants show increasing secretory activity with time. The SMG autografts remain functionally viable due to survival of parasympathetic ganglia within the gland and sympathetic reinnervation in transplanted gland tissue [10].

In contrast to the parasympathetic nerves, the cell bodies of the sympathetic nerves are located in the superior cervical ganglion in the neck, and postganglionic fibers innervate the salivary glands along the blood vessels, which branch from the carotid plexus of the external carotid artery [8]. Sympathetic innervation is important for salivary secretion and influences local inflammation [11, 12]. While it is clear that the autonomic nervous system is essential for adult salivary gland secretion [13], only more recently has it been shown how parasympathetic innervation is critical for the biogenesis of mouse salivary glands [14]. The interactions between the epithelium and nerves studied in mouse salivary glands may have implications for the regeneration of human salivary glands [13–16].

A description of salivary gland innervation also requires mention of the neurotransmitters that are involved in neuronal function. The neurotransmitter acetylcholine signals at the synapse of pre- and post-ganglionic parasympathetic and preganglionic sympathetic nerves and activates muscarinic receptors in the salivary glands via postganglionic parasympathetic fibers. Noradrenaline from postganglionic sympathetic fibers activates adrenergic receptors. Other neurotransmitters have roles in salivary gland function, such as vasoactive intestinal peptide, encephalin, substance P, neuropeptide Y, neurokinin A, pituitary adenylate cyclase-activating peptide, neuronal nitric oxide synthase and calcitonin gene-related peptide [17, 18], but it is not known if they affect gland biogenesis.

Salivary Gland Biogenesis

The biogenesis of human salivary glands has been described from histological reports and was recently reviewed [19]. Briefly, the development of the major salivary glands in humans begins at around 6–8 weeks of gestation. The placodes of the major glands start as thickenings of the oral ectoderm. Salivary gland biogenesis is characterized by branching morphogenesis of epithelium, which is closely associated with the developing vasculature and nerves to form a branched glandular structure of ducts with terminal buds that become acini by around 14 weeks. The neural crest-derived mesenchyme provides growth factors and other important molecular cues for epithelial branching morphogenesis. By 13–16 weeks in humans, the SMG appears well differentiated, with desmosome and microvillus projections from cells adjacent to the lumens. The basal lamina surrounds the epithelium, with a few elongated cells that appear similar to myoepithelial cells. The striated and intercalated ducts can be recognized as early as 16 weeks, with the acinar cells beginning to predominate the tissue by 20–24 weeks. In humans, the salivary glands continue to develop up to 28 weeks, at which stage secretory products can be seen in acini. At birth the glands are functional to secrete saliva [19].

In humans there are genetic diseases affecting salivary glands that inform us about salivary gland biogenesis. Similar gene mutations have been generated in mice to learn more about the genetic and cellular mechanisms of gland development. For example, patients with hypohidrotic ectodermal dysplasia present clinically with defects in the teeth, hair, sweat glands and salivary glands. Hypohidrotic ectodermal dysplasia is caused by mutations in ectodysplasin-A, its receptor EDAR or EDARRAD, an intracellular signaling molecule [20]. These genes make proteins that function together during gland development and are critical for signaling between the salivary epithelium and mesenchyme. Importantly, genetic manipulation of the mouse genome enables the development of models to study the molecular mechanisms of hypohidrotic ectodermal dysplasia. Mice lacking EDAR have SMG aplasia or hypoplasia, which is caused by reduced SMG epithelial cell proliferation, lumen formation and histodifferentiation [21, 22]. Other examples of genetic mutations in humans that cause problems with salivary biogenesis include patients with mutations in genes that affect fibroblast growth factor (FGF) signaling. Patients with a mutation or deletion of FGF10 have a syndrome called aplasia of lacrimal and salivary glands (OMIM180920). This syndrome results in salivary gland aplasia or hypoplasia, and the oral symptoms which present in childhood in-

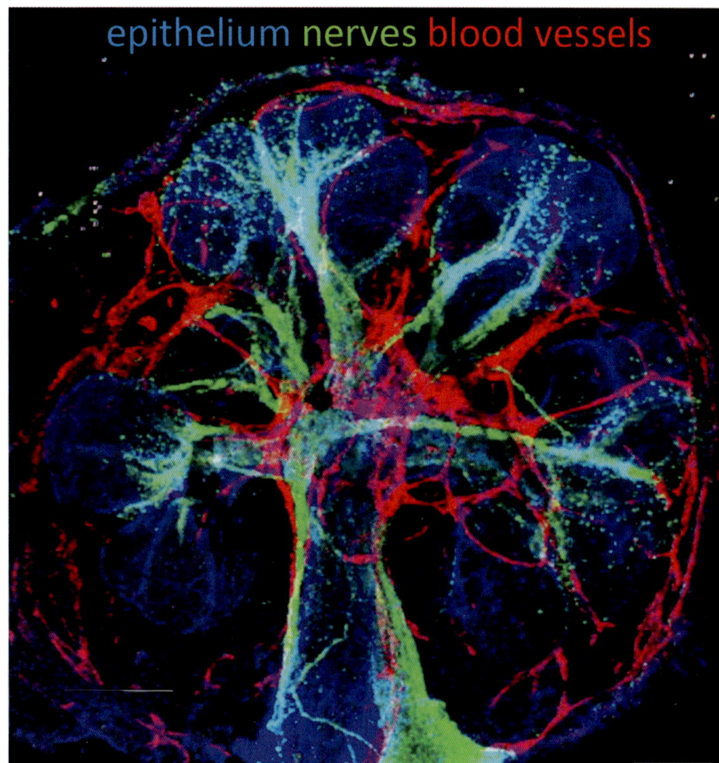

Fig. 3. There is a close association of the epithelium, nerves and blood vessels during SMG biogenesis. The projection of confocal sections shows immunostaining of the epithelium (blue), parasympathetic nerves (green) and blood vessels (red) of an E13 mouse SMG. The parasympathetic nerves are important for salivary gland biogenesis and regeneration.

clude xerostomia, with increased dental erosion, caries, periodontal disease and oral infections [23]. A similar condition occurs in lacrimoauriculodentodigital (OMIM 149730) syndrome, which occurs due to mutations in FGF receptors and/or FGF10. This syndrome presents as aplasia and/or hypoplasia of the salivary and lacrimal glands, and also abnormalities of the ears, eyes, face, mouth, teeth, digits and genitourinary system. Again, mice with similar genetic mutations have similar defects in salivary gland biogenesis and have been used to study how FGF signaling affects gland biogenesis [24, 25].

In addition to mouse genetic studies, ex vivo culture of embryonic mouse SMGs is an important tool to investigate how cell and extracellular matrix (ECM) interactions coordinate salivary gland formation and function. The reader is directed to a number of detailed reviews on gland development [16, 21, 26]. Similar to human SMGs, the mouse SMG begins with the formation of a placode in the oral epithelium at embryonic day 11.5 (E11.5), which enlarges and invaginates into the surrounding mesenchyme alongside nerves and blood vessels (fig. 3). The signals that spatiotemporally regulate placode initiation in mouse salivary glands are not known. However, studies in *Drosophila* suggest that global patterning genes, such as scr, select the initiating site location [27, 28]. By E12, a primary end bud distal to the primary duct is formed, and branching morphogenesis begins on E13 with repeating rounds of proliferation and clefting of the end buds and formation of secondary ducts. The repetitive branching continues with duct lumen formation occurring, and by E16, the end bud cells start to polarize and acinar differentiation begins. The SMG is competent to secrete saliva at birth [16, 21, 26].

Salivary gland stem cells play an important role in the SMG biogenesis. A stem cell is defined as a cell capable of unlimited self-renewal and differentiation into all cell types of an adult gland. Stem cells differentiate along a lineage into more committed progenitor cells that lose their self-renewing ability and eventually become terminally differentiated. They have a high potential for proliferation and give rise to different cell types of a specific lineage. The difference between stem and progenitor cells in the SMG is currently not well defined, so here we use the term stem/progenitor cells. During mitosis, stem/progenitor cells divide to produce either 2 undifferentiated daughter clones, 2 more committed cells, or 1 undifferentiated clone and 1 more committed cell. Maintenance refers to cell division that gives rise to at least 1 daughter stem/progenitor cell so that the number of stem/progenitor cells is maintained with successive generations. During salivary gland biogenesis, stem/progenitor epithelial cell fates are defined and modulated by communication with the surrounding microenvironment or niche, which includes mesenchyme, blood vessels and nerves. Stem/progenitor cells respond to signaling factors, direct cell-cell contact, and by binding to niche-derived ECM.

The mesenchymal niche is instructive for the epithelial stem/progenitor cells during branching morphogenesis. Classic developmental biology experiments used ex vivo embryonic tissue recombination to show how tissue-specific mesenchyme provides instructive cues to guide the patterning and morphogenic potential of epithelial progenitor cells [29, 30]. Multiple molecules, including components of the ECM, cell adhesion receptors, proteases and growth factors, mediate these instructive interactions. For example, epithelia from lung, mammary, and pituitary gland develop into structures mimicking salivary glands when recombined with E13 salivary mesenchyme [31–34]. The converse is not true for SMG embryonic epithelia, whose growth was only induced by urogenital or SMG mesenchyme.

The role of the endothelial niche, i.e. blood vessels, in salivary gland biogenesis remains to be investigated, but is likely important, considering the essential endothelial-epithelial interactions in the organogenesis of the liver and pancreas [35, 36]. Interstitial and immune cells, such as macrophages, may also have a role in gland biogenesis as has been shown for mammary gland development [37].

A major research goal is to identify and characterize the epithelial salivary gland stem/progenitor cells. Their location in adult glands is often described in relation to the main epithelial compartments of the gland: the ductal, acinar or myoepithelial cells. Stem/progenitor cells were initially identified as label-retaining cells because they were slowly dividing cells in the gland. In these experiments cells were pulse labeled with reagents that bind to DNA, and after months of continued growth only slowly dividing cells in the gland retained the DNA label. Initial studies showed that the label-retaining cells reside in the intercalated ducts and played a role in the regeneration of the gland after injury [38, 39]. Many studies have identified markers that can be used to isolate or label stem/progenitor cell populations from adult mouse glands, including CD49f (α_6-integrin), the tyrosine kinase receptor Kit, the transcription factor Ascl3, and the intracellular cytokeratin 5 (K5) [16, 40–42]. Further study of the Kit-expressing cells also showed that this cell population is located in the ducts of the adult SMG [43]. In addition, stem/progenitor cells of the ducts that were activated in response to injury were shown to express markers Sca1 and Kit [44, 45], $\alpha_6\beta_1$-integrin [46] or CD49f (α_6-integrin) and Thy1 [47, 48]. Furthermore, Ascl3, a transcription factor, also marks a stem/progenitor population that is localized in the salivary ducts [49]. Interestingly, other stem/progenitor cells compensate for the genetic ablation of the Ascl3-expressing stem cell population in the SMG. One interpretation of this data is that multiple progenitor populations exist during biogenesis that can compensate for the absence of each other, although further work is required to support this hypothesis.

The cytokeratins have been used extensively to study stem/progenitor cells in many epithelial organs. Basal cells expressing K5 and K14 (K5+K14+) label a population of stem/progenitor cells across different tissues, such as prostate, mammary gland and skin [50]. However in the salivary gland it appears these cytokeratins may label distinct progenitor cell populations. K5+ cells, which are mainly in the ductal structures, were shown to be a progenitor population in SMGs by genetic lineage tracing [14]. However, a significant finding was that the neuronal niche, i.e. the submandibular parasympathetic ganglion, which is present during SMG biogenesis, plays a critical role in the maintenance of these K5+ progenitors. Parasympathetic innervation and signaling via muscarinic receptors in the salivary epithelium, in combination with epithelial growth factor receptor signaling, maintained the K5+ cells during development and was critical for salivary gland biogenesis. More recently, a separate population of K14+ cells was shown to contain multipotent progenitors in SMGs [51]. The K14+ progenitors increase in cell number in response to Fgfr2b and Kit signaling and are located in the salivary gland end buds during biogenesis. Furthermore, epithelial Kit+ cells in the end buds produce neurotrophic factors, such as neurturin (Nrtn), which promote parasympathetic nerve survival and axon extension [15]. These parasympathetic nerves produce acetylcholine and further signal to the K5+ cells in the duct to continue to grow and differentiate. In sum, the different epithelial cells and their niches coordinate branching morphogenesis and gland biogenesis [14, 15]. These data have significant therapeutic implications in terms of repairing or regenerating adult salivary tissue.

Regeneration of Salivary Glands

The goal of regenerative medicine is to restore gland function in patients who suffer from irreversible loss of salivary gland function following resection of salivary tumors and therapeutic radiation of head and neck cancers. Additionally, patients suffer from gland hypofunction in diseases affecting salivary glands, such as Sjögren's syndrome, which is an autoimmune disease that results in inflammatory-mediated damage [52, 53]. A major research focus has been to regenerate salivary glands after radiation damage, because head and neck carcinoma is the sixth most common cancer in the world and affects tens of thousands of new patients every year. With irradiation being a primary treatment option, and the salivary glands often lying in the field of radiation, the subsequent salivary gland damage and xerostomia, or dry mouth, are significant causes of morbidity in these patients [54]. Modern advances in radiation therapy, such as intensity-modulated radiotherapy, allow for organ-sparing techniques; however, there are insufficient randomized clinical trials on intensity-modulated radiotherapy and no data to show that intensity-modulated radiotherapy reduces SMG radiation dosage [55]. Current treatment options for radiation-induced xerostomia include the use of saliva substitutes or parasympathetic agonists, such as pilocarpine, to stimulate salivary flow. However, there are problems with side effects of systemic parasympathomimetics, and along with their variable efficacy, this means more permanent treatment options are required. Clinical trials in patients treated with radiation therapy studied the effect of amifostine, an oxygen free radical scavenger, on postirradiation salivary gland damage. They showed a reduction in late xerostomia in patients on amifostine [56]. Also, in studies in rats, preirradiation stimulation of muscarinic acetylcholine receptors reduced parotid gland damage [57].

The study of mouse salivary glands provides mechanistic insight for developing new therapies for human salivary gland hypofunction. Salivary gland regeneration has the potential to permanently restore salivary gland secretory function in

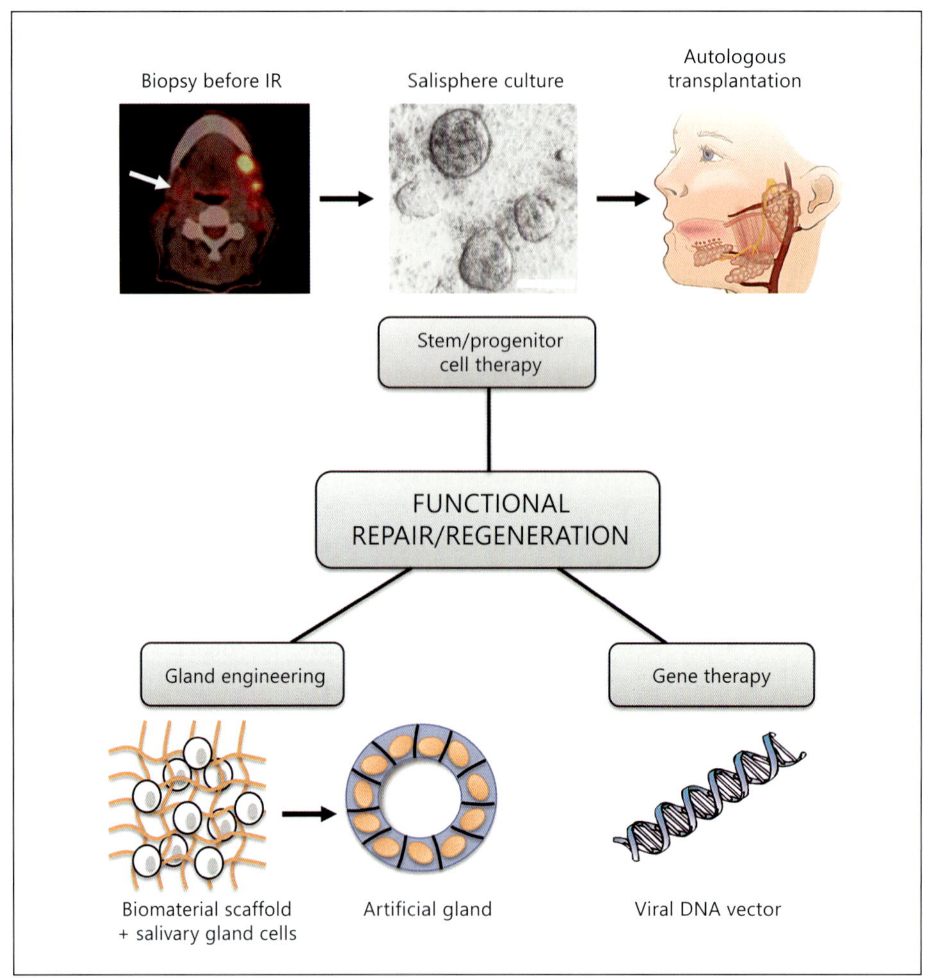

Fig. 4. Overview of three approaches for salivary gland repair or regeneration: stem/progenitor cell therapy, gene therapy and gland bioengineering. IR = Irradiation. For stem/progenitor cell therapy, a preoperative PET/CT scan, adapted from Nahmias et al. [77], shows the region of the unaffected SMG (arrow) used for a biopsy and the affected submandibular lymph nodes are yellow. Human salispheres are cultured from the biopsy; scale bar = 100 µm. Autologous transplantation of the spheres into the affected salivary gland occurs after therapy. Gland bioengineering involves combining biomaterials with cells to grow an artificial gland, which can be transplanted into a patient. Gene therapy uses viral vectors to transduce genes that repair or improve the secretory function of the gland.

patients with hyposalivation to improve their oral health and quality of life. The three main approaches that have been proposed are (1) gene therapy, by using viral vectors, (2) stem/progenitor cell-based therapy and (3) replacement with a bioengineered gland (fig. 4) [58]. In the following sections we will review these regenerative approaches.

Recently, pioneering gene therapy trials were completed in patients suffering from radiation-induced salivary hypofunction. The technique used the retroductal injection of an adenovirus that expressed a water channel (AQP1) to transduce the remaining ductal epithelium to secrete fluid. This trial was the culmination of many years of preclinical studies of gene transfer using rodents, minipigs and primates to develop the technique before translation to human clinical trials [59–62]. Aquaporins are water channels important for transcellular water transport in salivary glands, and human AQP1 gene transfer was shown to be efficacious in nonhuman in vivo models [61]. Recent results from the phase I clinical trial showed no deaths, serious adverse events or dose-limiting toxicities. Objective and subjective restoration of salivary gland function was reported in some of the patients who received the AdhAQP1 vector [60]. Future studies to improve viral vectors and optimize the vector dose may improve the clinical outcome. However, the effect of irradiation on the microenvironment or niche, including the extracellular matrix, endothelial, neuronal and stromal cells, is not fully understood. Complete functional regeneration requires restoration of all tissue types, which may not occur by gene therapy alone.

Cell-based therapies could ideally use a patient's own cells to regenerate tissues in organ systems. A common example in patients today is bone marrow allotransplants to replace and regenerate the hematopoietic system in patients with diseases such as aplastic anemia or leukemia. This technique was also used by the Coppes group to treat salivary gland hypofunction in mice [63]. They used bone marrow stem cells to regenerate irradiated mouse SMGs and showed an increase in salivary secretion after mobilization of bone marrow stem cells in the irradiated tissue. The transplanted cells were not found to be part of the epithelial compartment of the gland, but potentially secreted a paracrine acting factor that stimulated regeneration of the epithelia [63]. Consequently, others have taken this approach by showing that a potential proregenerative paracrine acting factor can be derived from bone marrow cells. They demonstrated this paracrine effect by

using a bone marrow stem cell supernatant or 'bone marrow soup', which was as effective as whole live bone marrow cells for repairing irradiated salivary glands [64]. They showed that the bone marrow-derived factors not only restored secretory function, but also protected cells, increased vascularity, and upregulated genes important for regeneration, e.g. genes for BMP7, EGF, NGF, MMP2, and cyclin D_1. These data lead to the logical suggestion that paracrine factors may be able to repair or stimulate the irradiated niche to stimulate surviving salivary gland stem/progenitor cells to regenerate functional salivary tissue.

A major advance in the field came from the Coppes laboratory which showed that the transplantation of adult mouse Kit+ cells resulted in functional regeneration of the salivary gland epithelium, which demonstrated that cell therapy in irradiated SMGs was a viable approach for regeneration [65, 66]. They have also shown that human salivary glands have cells with the same regenerative capacity, expressing Kit and capable of in vitro differentiation and self-renewal [67]. Autologous transplantation would involve a biopsy from a patient's own salivary gland before irradiation therapy, being used to culture, expand and maintain the stem/progenitor cells, which would be transplanted into the salivary gland stroma after radiation treatment. This model is the basis for ongoing studies to better understand adult salivary gland stem/progenitor cell populations and the signaling pathways important for maintaining their cell fate and expanding their number in culture.

It was previously shown that autonomic innervation of the salivary gland is required for salivary gland repair after atrophy [13]. In line with this reasoning, the role of parasympathetic innervation maintaining K5 progenitor cells during SMG biogenesis [14] leads one to ask what factors support the nerves during development, and whether these factors could be used to repair and support the function of nerves after irradiation damage. Nrtn is a neurotrophic factor expressed by salivary epithelium that binds to its receptor, GFR-α_2, on parasympathetic nerves to signal via RET, a tyrosine kinase coreceptor, and Src kinase [15]. Nrtn was shown to reduce neuronal apoptosis after irradiation, restore parasympathetic function and improve epithelial regeneration in a mouse ex vivo model [15]. Furthermore, adult human salivary glands damaged by irradiation were analyzed and shown to have reduced parasympathetic innervation. Therefore, it is possible that neurotrophic factors such as Nrtn will protect the parasympathetic nerves from irradiation damage and subsequently improve epithelial regeneration.

The concept of providing factors to protect or support the neuronal niche would also suggest that further investigations of sympathetic signaling and the role of Schwann cells are warranted to guide regeneration. Sympathetic nerve function was retained after irradiation in rats, suggesting irradiation affected the acinar response to stimulation, but not the sympathetic nerves themselves [68]. The difference in how irradiation affects sympathetic and parasympathetic nerves is not well understood, and in fact it may be the balance between the two that is important. Furthermore, Schwann cells have been shown to play critical roles in nerve development and survival [69] and may be important to protect nerves from irradiation damage. Understanding how nerves are affected by irradiation and how they might instruct or direct stem/progenitor cells to regenerate gland architecture will be important for cell-based or gland engineering approaches to regeneration.

The tissue engineering field aims to use biomaterials and cells to replace missing or damaged tissues. The field has evolved to understand the important role of the ECM environment for proper tissue regeneration. The ECM provides cues for cell function, stem cell self-renewal and differentiation, and dynamic remodeling necessary for engineering of complex tissues [70].

Models to engineer functional salivary gland tissue for autotransplantation and functional regeneration have been proposed. Engineering of an 'artificial' salivary gland may involve the use of synthetic or ECM-derived 3-dimensional scaffolds onto which cells are seeded to form a polarized secretory epithelium [71, 72]. Three-dimensional synthetic and ECM-based scaffolds can be used to culture and differentiate tissue-specific progenitor cell populations into composite glandular structures, with acinar, ductal, nerve and vascular components.

ECM components, such as laminin and glycosaminoglycans, attached to nanofiber scaffolds are important for regulating salivary gland epithelial cell proliferation and polarity [73]. ECM-based biomaterials are important for facilitating and regulating repair and regeneration of salivary glands by promoting membrane polarization. Cultured acinar cell populations formed lobular structures mimicking intact glands with functional activity, marked by α-amylase and Aqp5 expression when cultured on Matrigel or a 3-dimensional hydrogel containing a perlecan-derived bioactive peptide [74]. Furthermore, these cultures form 3-dimensional spheroids in hyaluronic acid-based hydrogels after long-term culture and respond to neurotransmitters for salivary protein secretion [75]. The surface architecture of a scaffold also influences cell differentiation. By fabricating a surface micropatterned with 'craters' lined with nanofibers of polylactic-coglycolic acid that mimicked the basement membrane of salivary acini, it was demonstrated that increased crater curvature promoted differentiation of salivary gland cells [76].

Translating these bioengineering approaches to human clinical trials is the next goal. Following in the footsteps of the pioneering work to translate gene therapy to the clinic, cell-based and/or tissue engineering approaches will require validation in nonhuman models to demonstrate their safety and efficacy. The clinical success of salivary gland regeneration may also require support of the stroma, nerves, vasculature and immune system. Engineering and clinical/translational investigations must be informed by the advancing knowledge in stem cell biology and salivary gland development, requiring a multidisciplinary approach with biologists, engineers and clinicians. Salivary gland biology provides a unique system to study biogenesis and regeneration, which will hopefully one day restore salivary gland function in patients who suffer from xerostomia.

Acknowledgments

We would like to thank Dr. I.M. Lombaert and Dr. W.M. Knosp for discussions and critical reading of this paper and S.E. Kibbey for illustration assistance. K.V.H. and M.P.H. are supported by the Intramural Research Program of the National Institute of Dental and Craniofacial Research, NIH, and K.V.H. is supported by the NIH Medical Research Scholars Program.

References

1 Carlson GW: The salivary glands: embryology, anatomy, and surgical applications. Surg Clin North Am 2000;80:261–273.

2 Johns ME: Salivary-glands – anatomy and embryology. Otolaryngol Clin North Am 1977;10:261–271.

3 Ten Cate AR: Oral Histology: Development, Structure, and Function, ed 5. St Louis, Mosby, 1998.

4 Ferreira JN, Hoffman MP: Interactions between developing nerves and salivary glands. Organogenesis 2013;9:199–205.

5 Davis RA, Anson BJ, Budinger JM, Kurth LR: Surgical anatomy of the facial nerve and parotid gland based upon a study of 350 cervicofacial halves. Surg Gynecol Obstet 1956;102:385–412.

6 Frommer J: Human accessory parotid-gland – its incidence, nature, and significance. Oral Surg Oral Med Oral Pathol Oral Radiol Endodont 1977;43:671–676.

7 Ishizuka K, Oskutyte D, Satoh Y, Murakami T: Multi-source inputs converge on the superior salivatory nucleus neurons in anaesthetized rats. Auton Neurosci 2010;156:104–110.

8 Kahle W, Frotscher M: Color Atlas of Human Anatomy – Nervous System and Sensory Organs, ed 1. Stuttgart, Thieme, 2003.

9 Tosios KI, Nikolakis M, Prigkos AC, Diamanti S, Sklavounou A: Nerve cell bodies and small ganglia in the connective tissue stroma of human submandibular glands. Neurosci Lett 2010;475:53–55.

10 Geerling G, Garrett JR, Paterson KL, Sieg P, Collin JR, Carpenter GH, Hakim SG, Lauer I, Proctor GB: Innervation and secretory function of transplanted human submandibular salivary glands. Transplantation 2008;85:135–140.

11 Mathison R, Davison JS, Befus AD: Neuroendocrine regulation of inflammation and tissue repair by submandibular gland factors. Immunol Today 1994;15:527–532.

12 Savastano LE, Castro AE, Fitt MR, Rath MF, Romeo HE, Munoz EM: A standardized surgical technique for rat superior cervical ganglionectomy. J Neurosci Methods 2010;192:22–33.

13 Proctor GB, Carpenter GH: Regulation of salivary gland function by autonomic nerves. Auton Neurosci Basic Clin 2007;133:3–18.

14 Knox SM, Lombaert IMA, Reed X, Vitale-Cross L, Gutkind JS, Hoffman MP: Parasympathetic innervation maintains epithelial progenitor cells during salivary organogenesis. Science 2010;329:1645–1647.

15 Knox SM, Lombaert IMA, Haddox CL, Abrams SR, Cotrim A, Wilson AJ, Hoffman MP: Parasympathetic stimulation improves epithelial organ regeneration. Nat Commun 2013;4:1494.

16 Knosp WM, Knox SM, Hoffman MP: Salivary gland organogenesis. Wiley Interdisc Rev Dev Biol 2012;1:69–82.

17 Ekstrom J, Garrett JR, Mansson B, Rowley PS, Tobin G: Depletion of large dense-cored vesicles from parasympathetic nerve terminals in rat parotid glands after prolonged stimulation of the auriculotemporal nerve. Regul Pept 1989;25:61–67.

18 Ekstrom J: Role of nonadrenergic, noncholinergic autonomic transmitters in salivary glandular activities in vivo; in Garrett JR, Ekstrom J, Anderson LC (eds): Neural Mechanisms of Salivary Secretion. Basel, Karger, 1999, vol 6, pp 94–130.

19 Knox S, Hoffman MP: Salivary Gland Development. Ames, Blackwell Publications, 2008.

20 Mikkola ML: Molecular aspects of hypohidrotic ectodermal dysplasia. Am J Med Genet A 2009;149A:2031–2036.

21 Tucker AS: Salivary gland development. Semin Cell Dev Biol 2007;18:237–244.

22 Jaskoll T, Zhou YM, Trump G, Melnick M: Ectodysplasin receptor-mediated signaling is essential for embryonic submandibular salivary gland development. Anat Rec A Discov Mol Cell Evol Biol 2003;271A:322–331.

23 Entesarian M, Matsson H, Klar J, Bergendal B, Olson L, Arakaki R, Hayashi Y, Ohuchi H, Falahat B, Bolstad AI, Jonsson R, Wahren-Herlenius M, Dahl N: Mutations in the gene encoding fibroblast growth factor 10 are associated with aplasia of lacrimal and salivary glands. Nat Genet 2005;37:125–127.

24 De Moerlooze L, Spencer-Dene B, Revest J, Hajihosseini M, Rosewell I, Dickson C: An important role for the IIIb isoform of fibroblast growth factor receptor 2 (FGFR2) in mesenchymal-epithelial signalling during mouse organogenesis. Development 2000;127:483–492.

25 Jaskoll T, Abichaker G, Witcher D, Sala FG, Bellusci S, Hajihosseini MK, Melnick M: FGF10/FGFR2b signaling plays essential roles during in vivo embryonic submandibular salivary gland morphogenesis. BMC Dev Biol 2005;5:11.

26 Patel VN, Rebustini IT, Hoffman MP: Salivary gland branching morphogenesis. Differentiation 2006;74:349–364.

27 Kerman BE, Cheshire AM, Andrew DJ: From fate to function: the Drosophila trachea and salivary gland as models for tubulogenesis. Differentiation 2006;74:326–348.

28 Denny PC, Ball WD, Redman RS: Salivary glands: a paradigm for diversity of gland development. Crit Rev Oral Biol Med 1997;8:51–75.

29 Grobstein C: Trans-filter induction of tubules in mouse metanephrogenic mesenchyme. Exp Cell Res 1956;10:424–440.

30 Kusakabe M, Sakakura T, Sano M, Nishizuka Y: A pituitary-salivary mixed gland induced by tissue recombination of embryonic pituitary epithelium and embryonic submandibular gland mesenchyme in mice. Dev Biol 1985;110:382–391.

31 Kratochwil K: Organ specificity in mesenchymal induction demonstrated in the embryonic development of the mammary gland of the mouse. Dev Biol 1969;20:46–71.

32 Sakakura T, Nishizuka Y, Dawe CJ: Mesenchyme-dependent morphogenesis and epithelium-specific cytodifferentiation in mouse mammary gland. Science 1976;194:1439–1441.

33 Nogawa H: Determination of the curvature of epithelial cell mass by mesenchyme in branching morphogenesis of mouse salivary gland. J Embryol Exp Morphol 1983;73:221–232.

34 Nogawa H, Mizuno T: Mesenchymal control over elongating and branching morphogenesis in salivary gland development. J Embryol Exp Morphol 1981;66:209–221.

35 Loreto C, Caltabiano R, Musumeci G, Caltabiano C, Greco MG, Leonardi R: Hepatocyte growth factor receptor, c-met, in human embryo salivary glands. An immunohistochemical study. Anat Histol Embryol 2010;39:173–177.

36 Matsumoto K, Yoshitomi H, Rossant J, Zaret KS: Liver organogenesis promoted by endothelial cells prior to vascular function. Science 2001;294:559–563.

37 Wynn TA, Chawla A, Pollard JW: Macrophage biology in development, homeostasis and disease. Nature 2013;496:445–455.

38 Carpenter GH, Khosravani N, Ekstrom J, Osailan SM, Paterson KP, Proctor GB: Altered plasticity of the parasympathetic innervation in the recovering rat submandibular gland following extensive atrophy. Exp Physiol 2009;94:213–219.

39 Denny PC, Liu PX, Denny PA: Evidence of a phenotypically determined ductal cell lineage in mouse salivary glands. Anat Rec 1999;256:84–90.

40 Lombaert IMA, Hoffman MP: Epithelial stem/progenitor cells in the embryonic mouse submandibular gland; in Tucker AS, Miletich I (eds): Salivary Glands: Development, Adaptations and Disease. Front Oral Biol. Basel, Karger, 2010, vol 14, pp 90–106.

41 Lombaert IMA, Hoffman MP: Epithelial stem/progenitor cells in the embryonic mouse submandibular gland; in Tucker AS, Miletich I (eds): Salivary Glands: Development, Adaptations and Disease. Front Oral Biol. Basel, Karger, 2010, vol 14, pp 90–106.

42 Lombaert IMA, Hoffman MP: Stem cells in salivary gland development and regeneration; in Huang GTJ, Thesleff I (eds): Stem Cells in Craniofacial Development and Regeneration. Hoboken, Wiley & Sons, 2013, pp 271–284.

43 Coppes RP, Stokman MA: Stem cells and the repair of radiation-induced salivary gland damage. Oral Dis 2011;17:143–153.

44 Hisatomi Y, Okumura K, Nakamura K, Matsumoto S, Satoh A, Nagano K, Yamamoto T, Endo F: Flow cytometric isolation of endodermal progenitors from mouse salivary gland differentiate into hepatic and pancreatic lineages. Hepatology 2004;39:667–675.

45 Okumura K, Nakamura K, Hisatomi Y, Nagano K, Tanaka Y, Terada K, Sugiyama T, Umeyama K, Matsumoto K, Yamamoto T, Endo F: Salivary gland progenitor cells induced by duct ligation differentiate into hepatic and pancreatic lineages. Hepatology 2003;38:104–113.

46 David R, Shai E, Aframian DJ, Palmon A: Isolation and cultivation of integrin alpha(6)beta(1)-expressing salivary gland graft cells: a model for use with an artificial salivary gland. Tissue Eng A 2008;14:331–337.

47 Matsumoto S, Okumura K, Ogata A, Hisatomi Y, Sato A, Hattori K, Matsumoto M, Kaji Y, Takahashi M, Yamamoto T, Nakamura K, Endo F: Isolation of tissue progenitor cells from duct-ligated salivary glands of swine. Cloning Stem Cells 2007;9:176–190.

48 Sato A, Okumura K, Matsumoto S, Hattori K, Hattori S, Shinohara M, Endo F: Isolation, tissue localization, and cellular characterization of progenitors derived from adult human salivary glands. Cloning Stem Cells 2007;9:191–205.

49 Yoshida S, Ohbo K, Takakura A, Takebayashi H, Okada T, Abe K, Nabeshima Y: Sgn1, a basic helix-loop-helix transcription factor, delineates the salivary gland duct cell lineage in mice. Dev Biol 2001;240:517–530.

50 Purkis PE, Steel JB, Mackenzie IC, Nathrath WB, Leigh IM, Lane EB: Antibody markers of basal cells in complex epithelia. J Cell Sci 1990;97:39–50.

51 Lombaert IM, Abrams SR, Li L, Eswarakumar VP, Sethi AJ, Witt RL, Hoffman MP: Combined Kit and FGFR2b signaling regulates epithelial progenitor expansion during organogenesis. Stem Cell Rep 2013;1:604–619.

52 Kok MR, Yamano S, Lodde BM, Wang JH, Couwenhoven RI, Yakar S, Voutetakis A, Leroith D, Schmidt M, Afione S, Pillemer SR, Tsutsui MT, Tak PP, Chiorini JA, Baum BJ: Local adeno-associated virus-mediated interleukin 10 gene transfer has disease-modifying effects in a murine model of Sjögren's syndrome. Hum Gene Ther 2003;14:1605–1618.

53 Vitali C, Bombardieri S, Moutsopoulos HM, Balestrieri G, Bencivelli W, Bernstein RM, Bjerrum KB, Braga S, Coll JQ, Devita S, Drosos AA, Ehrenfeld M, Hatron PY, Hay EM, Isenberg DA, Janin A, Kalden JR, Kater L, Konttinen YT, Maddison PJ, Maini RN, Manthorpe R, Meyer O, Ostuni P, Pennec Y, Prause JU, Richards A, Sauvezie B, Schiodt M, Sciuto M, Scully C, Shoenfeld Y, Skopouli FN, Smolen JS, Snaith ML, Tishler M, Todesco S, Valesini G, Venables PJW, Wattiaux EJ, Youinou P: Preliminary criteria for the classification of Sjögren's syndrome – results of a prospective concerted action supported by the European Community. Arthritis Rheum 1993;36:340–347.

54 Dirix P, Nuyts S, Van den Bogaert W: Radiation-induced xerostomia in patients with head and neck cancer – a literature review. Cancer 2006;107:2525–2534.

55 O'Sullivan B, Rumble RB, Warde P: Intensity-modulated radiotherapy in the treatment of head and neck cancer. Clin Oncol 2012;24:474–487.

56 Epstein JB, Thariat J, Bensadoun R-J, Barasch A, Murphy BA, Kolnick L, Popplewell L, Maghami E: Oral complications of cancer and cancer therapy. Cancer J Clin 2012;62:400–422.

57 Coppes RP, Zeilstra LJW, Kampinga HH, Konings AWT: Early to late sparing of radiation damage to the parotid gland by adrenergic and muscarinic receptor agonists. Br J Cancer 2001;85:1055–1063.

58 Redman RS: On approaches to the functional restoration of salivary glands damaged by radiation therapy for head and neck cancer, with a review of related aspects of salivary gland morphology and development. Biotech Histochem 2008;83:103–130.

59 Mastrangeli A, O'Connell B, Aladib W, Fox PC, Baum BJ, Crystal RG: Direct in vivo adenovirus-mediated gene transfer to salivary glands. Am J Physiol 1994;266:G1146–G1155.

60 Baum BJ, Alevizos I, Zheng C, Cotrim AP, Liu S, McCullagh L, Goldsmith CM, Burbelo PD, Citrin DE, Mitchell JB, Nottingham LK, Rudy SF, Van Waes C, Whatley MA, Brahim JS, Chiorini JA, Danielides S, Turner RJ, Patronas NJ, Chen CC, Nikolov NP, Illei GG: Early responses to adenoviral-mediated transfer of the aquaporin-1 cDNA for radiation-induced salivary hypofunction. Proc Natl Acad Sci 2012;109:19403–19407.

61 Shan Z, Li J, Zheng C, Liu X, Fan Z, Zhang C, Goldsmith CM, Wellner RB, Baum BJ, Wang S: Increased fluid secretion after adenoviral-mediated transfer of the human aquaporin-1 cDNA to irradiated miniature pig parotid glands. Mol Ther 2005;11:444–451.

62 Zheng C, Cotrim AP, Rowzee A, Swaim W, Sowers A, Mitchell JB, Baum BJ: Prevention of radiation-induced salivary hypofunction following HKGF gene delivery to murine submandibular glands. Clin Cancer Res 2011;17:2842–2851.

63 Lombaert IM, Wierenga PK, Kok T, Kampinga HH, de Haan G, Coppes RP: Mobilization of bone marrow stem cells by granulocyte colony-stimulating factor ameliorates radiation-induced damage to salivary glands. Clin Cancer Res 2006;12:1804–1812.

64 Tran SD, Liu Y, Xia D, Maria OM, Khalili S, Wang RW-J, Quan V-H, Hu S, Seuntjens J: Paracrine effects of bone marrow soup restore organ function, regeneration, and repair in salivary glands damaged by irradiation. PloS One 2013;8:e61632.

65 Lombaert IMA, Brunsting JF, Wierenga PK, Faber H, Stokman MA, Kok T, Visser WH, Kampinga HH, de Haan G, Coppes RP: Rescue of salivary gland function after stem cell transplantation in irradiated glands. Plos One 2008;3:e2063.

66 Nanduri LSY, Maimets M, Pringle SA, van der Zwaag M, van Os RP, Coppes RP: Regeneration of irradiated salivary glands with stem cell marker expressing cells. Radiother Oncol 2011;99:367–372.

67 Feng JL, van der Zwaag M, Stokman MA, van Os R, Coppes RP: Isolation and characterization of human salivary gland cells for stem cell transplantation to reduce radiation-induced hyposalivation. Radiother Oncol 2009;92:466–471.

68 Kohn WG, Grossman E, Fox PC, Armando I, Goldstein DS, Baum BJ: Effect of ionizing-radiation on sympathetic-nerve function in rat parotid-glands. J Oral Pathol Med 1992;21:134–137.

69 Yamaguchi Y, Yonemura S, Takada S: Grainyhead-related transcription factor is required for duct maturation in the salivary gland and the kidney of the mouse. Development 2006;133:4737–4748.

70 Daley WP, Peters SB, Larsen M: Extracellular matrix dynamics in development and regenerative medicine. J Cell Sci 2008;121:255–264.

71 Tran SD, Sugito T, Dipasquale G, Cotrim AP, Bandyopadhyay BC, Riddle K, Mooney D, Kok MR, Chiorini JA, Baum BJ: Re-engineering primary epithelial cells from rhesus monkey parotid glands for use in developing an artificial salivary gland. Tissue Eng 2006;12:2939–2948.

72 Sun T, Zhu J, Yang X, Wang S: Growth of miniature pig parotid cells on biomaterials in vitro. Arch Oral Biol 2006;51:351–358.

73 Cantara SI, Soscia DA, Sequeira SJ, Jean-Gilles RP, Castracane J, Larsen M: Selective functionalization of nanofiber scaffolds to regulate salivary gland epithelial cell proliferation and polarity. Biomaterials 2012;33:8372–8382.

74 Pradhan S, Zhang C, Jia X, Carson DD, Witt R, Farach-Carson MC: Perlecan domain IV peptide stimulates salivary gland cell assembly in vitro. Tissue Eng A 2009;15:3309–3320.

75 Pradhan-Bhatt S, Harrington DA, Duncan RL, Jia X, Witt RL, Farach-Carson MCP: Implantable three-dimensional salivary spheroid assemblies demonstrate fluid and protein secretory responses to neurotransmitters. Tissue Eng A 2013;26:26.

76 Soscia DA, Sequeira SJ, Schramm RA, Jayarathanam K, Cantara SI, Larsen M, Castracane J: Salivary gland cell differentiation and organization on micropatterned PLGA nanofiber craters. Biomaterials 2013;34:6773–6784.

77 Nahmias C, Carlson ER, Duncan LD, Blodgett TM, Kennedy J, Long MJ, Carr C, Hubner KF, Townsend DW: Positron emission tomography/computerized tomography (PET/CT) scanning for preoperative staging of patients with oral/head and neck cancer. J Oral Maxillofac Surg 2007;65:2524–2535.

Matthew P. Hoffman
NIDCR, NIH, Bldg 30/Rm 433
30 Convent Drive, MSC 4370
Bethesda, MD 20892-4370 (USA)
E-Mail mhoffman@mail.nih.gov

Ligtenberg AJM, Veerman ECI (eds): Saliva: Secretion and Functions.
Monogr Oral Sci. Basel, Karger, 2014, vol 24, pp 14–29 (DOI: 10.1159/000358781)

Salivary Secretion: Mechanism and Neural Regulation

Gordon B. Proctor · Guy H. Carpenter

Salivary Research Unit, King's College London Dental Institute, London, UK

Abstract

Maintenance of a film of saliva on oral surfaces is dependent upon nerve-mediated, reflex salivary gland secretion. Afferent signalling arises from taste, olfaction and mastication and is modified by signalling from other centres in the central nervous system before efferent signals are delivered to salivary glands in autonomic nerves. Salivary fluid secretion is largely dependent upon cholinergic signalling from parasympathetic nerves whilst the protein content of saliva is additionally dependent upon signalling by neuropeptides and, in the major (parotid, submandibular and sublingual) salivary glands, by sympathetic nerves and the release of noradrenaline. There have been significant recent advances in our understanding of the membrane transport proteins involved in intracellular calcium signalling in salivary acinar cells in response to nerve stimulation and of the ion transport proteins responsible for acinar cell secretion of saliva. Salivary glands retain an ability to regenerate following extreme atrophy, and autonomic nerves have an important role in both gland development and maintenance of long-term normal function. Continued advances in the understanding of the nerve-mediated regulation of salivary glands should help in the development of strategies for preventing chronic oral dryness resulting from drugs or atrophic disease associated with inflammation and irradiation.

Salivary glands fulfil a huge range of functions in different species, and even amongst mammals there is great variety in salivary gland morphology and the control of salivation by nerves, reflecting adaptation to diet and environment [1]. In man, the paired major salivary glands, parotid, submandibular and sublingual, along with hundreds of small, minor submucosal salivary glands provide a film of mixed saliva that coats and protects the oral mucosal and tooth surfaces. Salivary secretion is maintained at a 'default' rate in man creating a mobile but slow-moving film and replenishing/replacing proteins adsorbed to the underlying soft and hard oral surfaces. Upon this 'default' secretion of 'unstimulated' or 'resting' saliva there is superimposed a secretion of much greater volumes of saliva in response to taste, smell and chewing during periods of food intake [2]. The term 'unstimulated' saliva is a convenient way to discriminate from a saliva secreted in response to an overt taste or chewing stimulus but is in some ways a misnomer since salivary secretion of fluid is only unstimulated in the complete absence of neural activation, which does not apply in the conscious subject with an intact innervation.

Resting whole-mouth saliva is subject to a circadian rhythm in flow rate and salt content reach-

Table 1. Contribution of different salivary glands to the volume of whole-mouth saliva (WMS)

	Resting, ml/min	Resting, %	Stimulated, ml/min	Stimulated, %
WMS	0.35	100	2.0	100
Parotid glands	0.1	28	1.05	53
Submandibular/sublingual glands	0.24	68	0.92	46
Minor glands	<0.05	4	<0.1	1

Figures are based on data from Kalk et al. [104].

ing a peak flow in the mid-afternoon (around 15.00 h) and a nadir in the early morning (around 3.00 h); the salt concentration is inversely related to the flow rate [3].

The properties and composition of mixed saliva delivered to the mouth at rest differ from that secreted during eating, reflecting altered contributions from each of the salivary glands. Parotid gland acinar cells do not produce visco-elastic mucin glycoproteins and make a relatively greater contribution to whole-mouth saliva during stimulation (table 1), whilst the contribution of acinar cells from other salivary glands, most of which secrete mucin, is relatively greater under resting secretions. This difference in saliva quality and composition during rest compared to feeding enables saliva to fulfil the different functional requirements of immune exclusion and protection and food processing to swallowing.

Reflex Secretion of Saliva

Afferent Mechanisms
Secretion from the major salivary glands is evoked by interaction of tastants with different receptors on taste buds located predominantly in the epithelium on the dorsum of the tongue and following activation of mechanoreceptors in the periodontal ligament and mucosae [4]. Minor salivary glands may also increase secretion in response to taste stimulation [5] but perhaps movement and tactile stimulation of the mucosa play a more important role in labial and palatine minor glands [6, 7]. The submandibular and sublingual glands but not the parotid gland increase secretion in response to different smells associated with food [8]. The sensation of cold in the mouth can evoke a flow of saliva [9] and can increase salivation in response to liquid gustatory stimulation. Temperature, pungent substances such as capsaicin and hydroxyl-α-sanshool and cooling agents such as menthol activate TRP (transient receptor potential) channels; a range of these channels including TRPV1, -3, -4, TRPM8 and TRPA1 are expressed on trigeminal nerve endings, taste receptors and oral keratinocytes [10], and some have been shown to evoke salivary secretion [11, 12].

Central Integration of the Salivary Reflex
Taste, mechanical or pungency signals generate afferent signals in fibres of the facial (CNVII), glossopharyngeal (CNIX) and trigeminal (CNV) nerves. The nucleus of the solitary tract is innervated by the CNVII and CNIX and sends interneurons to the salivary centres, i.e. the superior and inferior salivary nuclei in the medulla oblongata. Interneurons presumably supply the primary sympathetic salivary centres which are located in the upper thoracic segments of the spinal cord although it remains unclear precisely where in this region [13, 14]. Efferent nerve fibres from the salivary nuclei conduct efferent signals via the

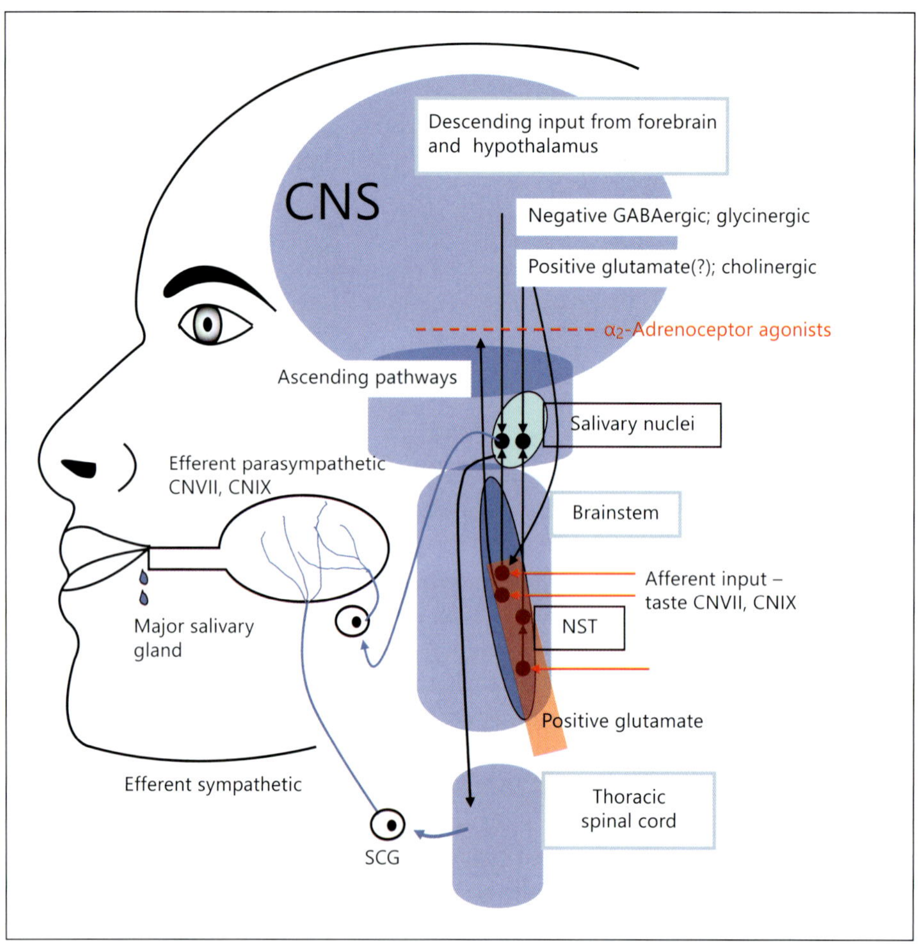

Fig. 1. Central connections influencing salivary gland reflex secretion. Afferent signals entering the central nervous system (CNS) at the nucleus of the solitary tract (NST) are relayed by interneurons to the salivary nuclei. Signals in descending neural pathways from higher cortical centres and hypothalamus to the nucleus of the solitary tract and salivary nuclei can enhance or reduce efferent stimulation of salivary gland secretion. α₂-Adrenoceptor agonists appear to modify descending signalling and reduce salivary secretion. Descending pathways from the salivary nuclei connect with efferent sympathetic nerves in the thoracic spinal cord at the level of the superior cervical ganglion (SCG).

chorda lingual nerve to the submandibular ganglion and thence to the submandibular and sublingual glands. The parotid gland is supplied by efferent fibres in the glossopharyngeal (tympanic branch) nerve to the otic ganglion and postganglionic fibres in the auriculotemporal nerve (fig. 1). There also appears to be a contribution to the parotid gland efferent supply from the facial nerve [15]. Minor salivary glands are supplied by parasympathetic nerve fibres in the buccal branch of the mandibular nerve, the lingual nerve and the palatine nerve.

The salivary reflex is profoundly influenced by central nerves from other nuclei in the brain supplying the salivary nuclei in the medulla oblongata. The salivary nuclei have various inputs from the frontal cortical areas as demonstrated by nerve-tracing experiments [16]. This central neural activity appears to contribute towards the resting rate of salivary secretion into the mouth since salivary flow rates are lower during sleep and virtually absent during anaesthesia. Retrograde labelling of neurons has demonstrated that the primary parasympathetic salivary centres form connections with the lateral hypothalamus where the regulation of feeding, drinking and body temperature occurs [14]. Both excitatory (γ-aminobutyric acid-containing) and inhibitory (glycine-containing) nerves synapse with the salivary centres [13]. Suppression of impulse traffic from the salivary nuclei to salivary glands leading to reduced salivation and dry mouth is most obviously demonstrated during fear and anxiety, and, like other autonomic regulation, involves a complex interaction with higher (limbic and cortical) centres in the brain. Different sensory modalities, including auditory, visual and somatosensory, are associated with fear and may potentially impact on salivary secretion through pathways in the amygdala, the hypothalamus and the brainstem.

Previous neuro-anatomical studies have also shown that there are also cholinergic inputs to the salivary centres from other nuclei including the substantia innominata, pedunculopontine nucleus and lateral dorsal tegmental nucleus. It has recently been demonstrated that neurons in the superior salivary nucleus express M_3 and other muscarinic acetylcholine receptors [17]. Since cholinergic neurons from the pedunculopontine nucleus and lateral dorsal tegmental nucleus are associated with maintenance of wakefulness and show increasing impulses during wakefulness, it may be that these inputs enhance the activity of superior salivary nucleus neurons and increase salivation during wakefulness whilst reduced impulse input from the pedunculopontine nucleus and lateral dorsal tegmental nucleus suppresses salivation during sleep. Salivary secretion is reduced during sleep, and it may be that these neural inputs to the salivary centres also account for the circadian pattern of resting or unstimulated salivation observed in man [3]. The presence of muscarinic receptors on neurons of the salivary nuclei may also partly explain the observed effects on salivary secretion evoked by intracerebroventricular injection of pilocarpine or atropine which were found to respectively stimulate and inhibit salivation [18, 19].

Significant advancements in our understanding of the brain have been made possible by functional MRI [20]. By the injection of labelled glucose (or other substrates) the active regions of the brain can be imaged when stimuli such as food or drinks are put in the mouth. Despite some recent advances in understanding of how tastes are perceived [21] relatively little attention has been paid as to how taste affects the salivary nuclei. Nerve recording in animals suggests that there are significant inputs from breathing and pulse activities [16]. However, some care must be paid towards using animal studies to speculate about functions in humans. For instance, most people believe that the thought of foods activates salivary secretion, the so-called mouth-watering [22]. However, neither Pavlov nor Lashley found any evidence to support the presence of a conditional salivary reflex in man. fMRI studies have demonstrated the considerable differences between animal and human brains in response to food [23]. Experiments by one of the authors (G.C.) also suggest that just the thought of food does not sustain a stimulated salivary flow [24] and that most mouth-watering experiences are the result of smells evoking submandibular/sublingual salivary flow [8]. Using flow meters it was possible to detect, particularly when subjects were hungry, small spikes of salivary flow. It was speculated that facial muscles compress the turgid ducts coming from salivary glands to the mouth to cause small transient 'flows' of saliva that can be easily perceived by the subject.

α_2-Adrenoceptor agonists (e.g. clonidine) and antagonists (e.g. yohimbine) have been demonstrated to act centrally in studies of reflex secretion in human subjects and cholinergically evoked secretion in animal models. α_2-Adrenoceptor blockade can increase salivary secretion whilst α_2-adrenoceptor agonists inhibit secretion [25, 26]. It appears that adrenergic agonists such as amphetamine exert an inhibitory effect on the flow of saliva through the release of noradrenaline from nerves in the medulla causing activation of inhibitory α_2-adrenoceptors rather than through a peripheral vasoconstrictive effect [27]. These central effects of amphetamine that cause a dry mouth contrast with its action in the periphery leading to increased secretion of protein by salivary cells and increased salivary protein concentration.

Efferent Autonomic Regulation of Salivary Secretion

Salivary gland cells are intimately associated with the autonomic nervous system, and over many years this relationship has fascinated some notable researchers from Claude Bernard in the 19th century to John Langley, Nils Emmelin and John Garrett at the end of the 20th century [28]. Parasympathetic and sympathetic nerves run together with Schwann cells to the target cells in salivary glands [29]. Other target cells are supplied by unmyelinated axons. Parasympathetic and sympathetic nerves are in contact with many cell types in salivary glands including acinar, ductal, myo-epithelial cells and blood vessels. The extent of innervation of salivary glands by sympathetic nerves varies greatly; the parotid and submandibular glands of the rat, mouse and man receive extensive sympathetic innervations whilst mucus-secreting glands such as the rat and human sublingual and the human minor salivary glands receive a sparse adrenergic innervation which appears to be directed to the vasculature rather than the parenchyma [30, 31]. In addition to the main neurotransmitters acetylcholine and

adrenaline there are a range of neuropeptides, including substance P and vaso-active intestinal peptide, within nerves in salivary glands [32]. Neuropeptide-containing nerves supply blood vessels and parenchymal cells and show distinct innervation patterns; for example, vaso-active intestinal peptide-containing nerves are more numerous around the mucous acinar cells in the human submandibular gland [33]. Some neuropeptides are also found in sensory nerve fibres around ducts and blood vessels within the salivary glands [12].

The acute control of salivary secretion and blood flow was demonstrated using animal models under anaesthesia and has been reviewed previously [34]. An assay of salivary protein concentration reveals that sympathetic nerve stimulation evokes a protein-rich secretion whilst parasympathetic stimulation evokes a larger volume of saliva. Dual nerve stimulation experiments have demonstrated that the individual actions of the nerves, particularly protein secretion evoked by the sympathetic nerve, are augmented in rat parotid [35] and submandibular glands [36, 37]. Such dual stimulation experiments are thought to better reflect the events leading to reflex secretion of saliva, since it is expected that both parasympathetic and sympathetic impulses are acting on secretory cells simultaneously. The paramount importance of an intact parasympathetic innervation is clear when one considers the dryness caused by blockade of the effects of acetylcholine by atropine and its analogues. Studies in man and the rat have demonstrated that sympathetic impulses make a contribution to the amount of protein secreted under reflex taste stimulation [38]. Although adrenergic signalling from sympathetic nerves leads to an augmentation of protein secretion by parotid and submandibular glands, mucin secretion from mucous glands such as the rat sublingual gland and human minor glands is dependent upon parasympathetic stimulation and peptidergic stimulation [39].

The Coupling of Autonomic Nerve Stimulation to Secretion

Coupling of Fluid Secretion

Salivary secretion is largely dependent upon the activation of muscarinic receptors on salivary acinar cells by acetylcholine released from parasympathetic nerves [34], M_3 acetylcholine receptors in the rat parotid gland [40] and both M_3 acetylcholine receptors and M_1 acetylcholine receptors in the submandibular gland [41, 42]. Acinar cell activation of fluid transport is achieved through increases in intracellular calcium concentration and binding of calcium to ion-transporting proteins. The acinar cell muscarinic receptors are G-protein-coupled receptors; binding of acetylcholine leads to a G-protein/phospholipase C-mediated generation of inositol triphosphate (IP_3) from phosphatidylinositol 4,5-bisphosphate. IP_3 interacts with IP_3 receptors (IP_3Rs) on the endoplasmic reticulum (ER) causing release of stored calcium [43]. The increase in cytoplasmic calcium originates in the apical region of acinar cells, where IP_3Rs are concentrated, and is propagated to other parts of the cell through calcium-induced activation of further calcium release via IP_3Rs. Cytoplasmic calcium levels are tightly controlled by rapid removal of calcium through the actions of plasma membrane and ER calcium pumps. Sustained salivary secretion requires influx of extracellular calcium across the plasma membrane of acinar cells referred to as store-operated calcium entry, and this is a research area where knowledge has greatly increased over the last 10 years [44, 45]. Store-operated calcium entry has been shown to be dependent upon the presence of 3 proteins, STIM1, Orai1 and TRPC1 channels. TRPC1 and Orai1 are membrane-bound channels whilst STIM1 is expressed on the ER. During stimulation of salivary secretion when there is depletion of the intracellular store of calcium, STIM1 translocates to the plasma membrane and forms complexes with TRPC1 and Orai1 leading to entry of extracellular calcium into the ER [46, 47]. Formation of the complexes of STIM1, Orai1 and TRPC1 appears to occur in regions of the plasma membrane termed caveolae which are cholesterol-enriched microdomains associated with receptors and ion channels [48]. Other receptors (α_1-adrenoceptor; substance P neurokinin 1 receptor; P2Y receptor; P2X receptors) utilize intracellular calcium signalling mechanisms but may make comparatively minor contributions to salivary fluid secretion under physiological conditions.

Coupling of Protein Secretion

Exocytosis of protein storage granules by salivary acinar cells is principally activated by noradrenaline release from sympathetic nerve endings binding to β_1-adrenoceptors and increases in G-protein-coupled adenylate cyclase activity with the generation of increased levels of intracellular cAMP [40]. Signalling from parasympathetic nerves can also give rise to substantial salivary protein secretion via release of vaso-active intestinal peptide [14] which also acts through increases in intracellular cAMP. However, cholinergic stimuli alone can give rise to the release of protein by a coupling mechanism independent of cAMP, involving elevated intracellular calcium and activation of protein kinase C [49].

Simultaneous activation of sympathetic and parasympathetic nerve supplies as occurs during reflex secretion, leads to 'augmented' secretion of amylase and other salivary proteins [50] and appears to reflect a 'cross-talk' between the intracellular calcium and cAMP secretory signalling pathways [51, 52]. The mechanism of cross-talk may involve a potentiation of the release of calcium due to phosphorylation of IP_3Rs by cAMP-dependent protein kinase A [53]. Acinar cells activated by cAMP in this way were found to elevate intracellular calcium levels in response to sub-threshold doses of methacholine. Denervation experiments in animal models have also revealed how the branches of the autonomic nervous system interact during coupling of nerve stimuli to secretion [54, 55].

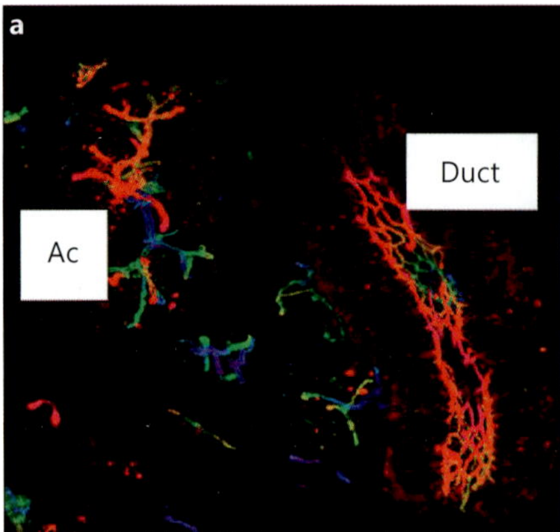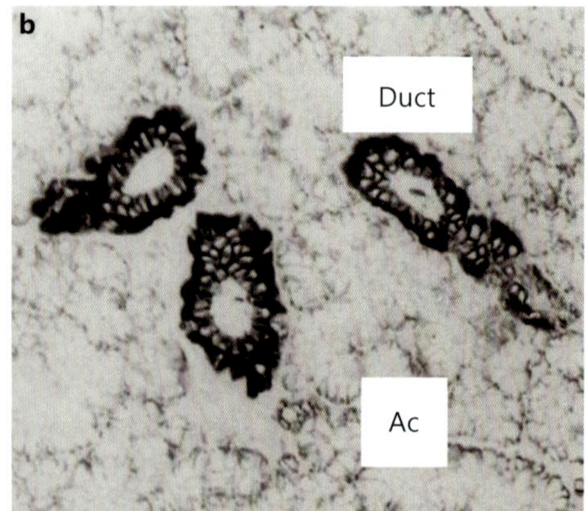

Fig. 2. Polarity of salivary cells. Ac = Acinar cell; Duct = ductal cell. **a** Immunofluorescence of ZO-1, a component of tight junctions located in the apical membrane of salivary acinar and ductal cells in the rat submandibular gland. **b** Immunohistochemistry of $Na^+/K^+/ATPase$ in the basolateral membranes of salivary acinar cells and striated ductal cells of the cat parotid gland. The striated ductal cells undertake relatively much greater amounts of ionic transport and show an intense staining.

Salivary Gland Secretory Mechanisms

Fluid and Electrolyte Secretion
The directional movement of salivary fluid and protein into the acinar lumina of salivary glands and to the mouth is dependent upon salivary acinar cell polarity, created by close interaction between adjacent cells with formation of tight junctions and maintained by interaction of the basal aspect of cells with basal laminae. Tight junctions are protein complexes formed principally from the transmembrane proteins claudins, occludins and junctional adhesion molecules. Tight junctions interact with zonula occludins which are intracellular scaffold proteins linked to the actin cytoskeleton of cells, and together the interaction of tight junctions, zonula occludins (fig. 2a) and the cytoskeleton creates a polarized epithelial layer which regulates transcellular and paracellular movement of salivary components [56]. Cells lining the ductal system of salivary glands are similarly polarized but in this case the tight junctions are watertight indicative of a greater number of tight junctional contacts between cells; similar differences in the leakiness of tight junctions are seen in different parts of the kidney tubular system.

Acinar cells secrete salivary fluid and there appears to be a minimal contribution to the overall volume of secretion by the ductal system through which saliva passes to the mouth. Salivary acinar epithelial cells are salt secreting, and it is the movement of salt across the epithelium from interstitial fluid into acinar lumina that leads to water movement and formation of salivary fluid (fig. 3). Secretion of saliva is ultimately dependent upon an increased activity of the $Na^+/K^+/ATPase$ located in the basolateral membrane of acinar cells (fig. 2b) and the maintenance of low intracellular Na concentrations relative to the extracellular environment. Inhibition of $Na^+/K^+/ATPase$ activity with ouabain inhibits salivary secretion [57]. In vitro electrophysiological studies utilizing specific inhibitors and ion-free or substituted buffers have provided data indicating the role of specific ion-transporting proteins in salivary secretion. Studies of salivary glands from different species indicate

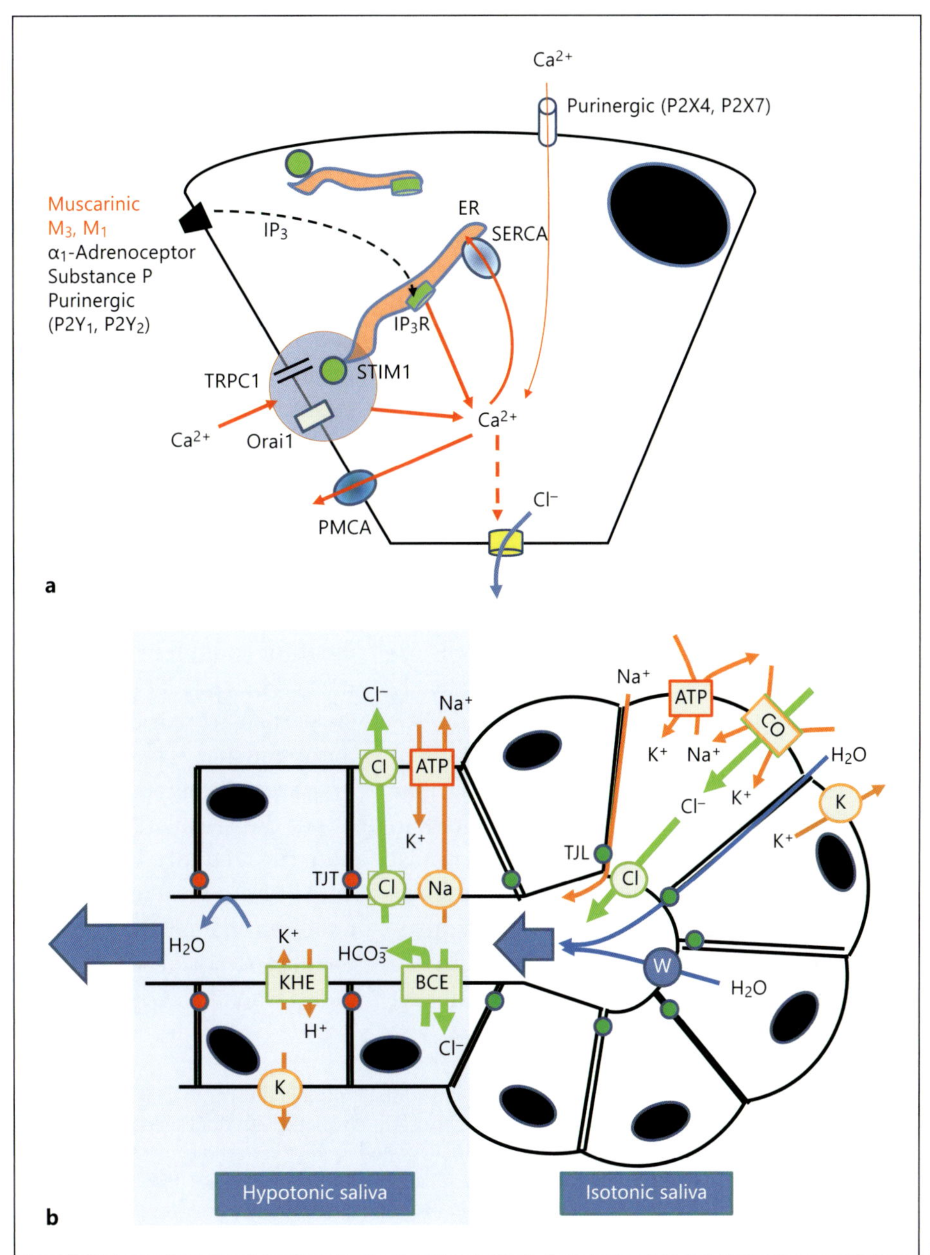

(*For legend see next page.*)

substantial variation in the details. Even amongst the most commonly studied laboratory models, the submandibular and parotid glands of the rabbit, rat and mouse, variation is seen in the impact of selective pharmacological inhibition of different transporters [58]. However, it is generally accepted that salivary secretion is dependent upon a calcium-activated chloride channel in the apical membrane of acinar cells crucial for salivary fluid secretion as shown by micropuncture techniques [59]. Recent studies by Melvin et al. [45], in which candidate transporters were heterologously expressed in HEK293 cells and knocked out in mice, have identified TMEM16A as the calcium-activated chloride channel, since efflux of chloride from mouse submandibular gland cells in vitro from knockout mice is abolished [60]. Movement of chloride from interstitial fluid into acinar cells across the basolateral membrane is largely abolished by inhibition or deletion of Nkcc1, a sodium-potassium-chloride cotransporter that utilizes the sodium gradient in cells to drive chloride influx [61]. Residual chloride movement following Nkcc1 deletion appears to depend upon AE2, a chloride/bicarbonate anion exchange transporter, and NHE1, a sodium proton exchange transporter, as demonstrated using selective inhibitors such as DIDS and amiloride and localization to acinar cells using specific antibodies immunocytochemically [58]. Following the movement of chloride, sodium moves along an electrochemical gradient by a paracellular route into acinar lumina. The osmotic gradient created by salt accumulation in acinar lumina leads to movement of water most likely by both paracellular and transcellular routes. Aquaporin 5 is expressed in apical membranes of acinar cells [62] and appears to play an important role in salivary fluid secretion since it

is decreased by 50% in aquaporin 5 knockout mice [63]. Water is therefore drawn into the ductal system either by flow through aquaporin channels or around cells and through the tight junctions, the leakiness of which is likely to be regulated (fig. 3).

Saliva entering the mouth from major salivary glands is hypotonic enabling the tasting of salt in food. Saliva secreted by acinar cells is isotonic, and as it flows through the ductal system of the major salivary glands salt is removed, principally by striated duct cells, and saliva is rendered hypotonic. The degree of hypotonicity is dependent upon the salivary flow rate; consequently stimulated saliva secreted at an increased flow rate has a higher salt concentration [64, 65]. The removal of sodium and chloride by ductal cells is again dependent upon creation of a transmembrane gradient for sodium by a basolaterally located sodium potassium ATPase. In fact striated duct cells are particularly enriched in this enzyme and, with the abundance of basolaterally located mitochondria, are well equipped to transport large amounts of salt transcellularly into the glandular interstitium (fig. 2b [66]). Entrance of sodium into ductal cells from the ductal lumen is dependent upon a sodium channel (ENaC) expressed in the apical membrane, since sodium absorption is greatly reduced when ENaC is knocked out. Inward move-

Fig. 3. Secretion of fluid by salivary cells. **a** Calcium signalling mechanisms in salivary acinar cells activate an apical chloride channel and initiate the secretion of salivary Cl⁻ and subsequent fluid secretion. Calcium is released from the ER by IP$_3$, a second messenger generated principally following activation of muscarinic receptors by acetylcholine. Raised cytoplasmic calcium concentrations are reduced by plasma membrane (PMCA) and ER (SERCA) calcium ATPases leading to depletion of the ER calcium store. In order to maintain increased intracellular calcium and a sustained secretory response, store-operated extracellular calcium entry is activated requiring the interaction of the ER membrane protein STIM1 with TRPC channels and Orai1 in the plasma membrane. **b** A summary of some of the membrane transporting proteins contributing to secretion and modification of saliva by salivary acinar and ductal cells. Chloride transport through acinar cells is achieved via uptake by a basolaterally located Na⁺, K⁺, 2Cl⁻ cotransporter (CO) and release through the calcium-activated apical chloride channel (Cl). The process is sustained by Na⁺/K⁺/ATPase (ATP). Sodium enters the acinar lumen paracellularly through leaky tight junctions (TJL), and water follows via aquaporin 5 (W) or paracellularly. Ductal cell removal of sodium from saliva is via an apical sodium channel (Na) and a basolateral Na⁺/K⁺/ATPase. Chloride removal from saliva occurs via different chloride channels in the apical and basolateral membranes of ductal cells. Removal of salt is not accompanied by water since ductal cell tight junctions (TJT) are not leaky and aquaporins are not expressed in apical membranes. Ductal cells can secrete bicarbonate and potassium via unidentified apical bicarbonate chloride (BCE) and potassium proton (KHE) exchangers.

ment of chloride shows some dependency on the expression of the cystic fibrosis transmembrane regulator, since it is reduced in submandibular glands of mice expressing the cystic fibrosis transmembrane regulator with the delta F508 mutation. The sodium potassium ATPase achieves removal of sodium across the basolateral membrane into the interstitial space of the gland. A chloride channel (CLCN2) is present in the basolateral membrane of ductal cells but has not been demonstrated to be required for removal of chloride from ductal cells [58, 60].

Bicarbonate is an important component of saliva since it plays a major role in buffering salivary pH near neutrality and preventing dissolution of tooth mineral, which increases in the presence of protons. Bicarbonate also facilitates solubilization of macromolecules and alters the solubility and rheological properties of mucins. Salivary acinar cells can secrete bicarbonate but it appears that ductal cells play the major role in bicarbonate secretion into saliva. The accumulation of bicarbonate in ductal cells is most likely achieved by a sodium-bicarbonate cotransporter (NBel-B) and a sodium proton exchanger (NHE1) located in the basolateral membrane. Entrance of bicarbonate into ductal saliva is achieved by an unidentified anion exchanger [58, 67]. Since the bicarbonate concentration of stimulated saliva is many times higher than that of unstimulated saliva, ductal bicarbonate secretion is most likely subject to control by autonomic nerve-mediated stimuli (fig. 3).

Thiocyanate, iodide, pertechnetate and nitrate are transported into saliva, and for different reasons each of these ions is significant in functional studies of saliva and salivary glands. Thiocyanate is converted to hypothiocyanate, a bacteriostatic molecule, in the presence of salivary peroxidase and bacterial derived hydrogen peroxide. Iodide similarly has bacteriostatic properties. Pertechnetate is important in imaging of salivary gland function, and nitrate can also have a bacteriostatic role when metabolized to nitrite but perhaps more significantly can have an influence on systemic blood pressure [68]. It appears that each of these anions competes for transportation into saliva [69] but each can reach concentrations above circulating levels; for example, approximately 25% of circulating nitrate is actively taken up by salivary glands and transported into saliva to reach levels over tenfold higher than those in serum. Salivary glands express a basolateral sodium-iodide cotransporter (NIS) in ductal cells that appears to be responsible for the uptake and accumulation of iodide and the other anions using the gradient of sodium concentration to drive anions into the cell. However, recently sialin, a sialic acid/proton cotransporter, has been identified as significant cotransporter of nitrate/protons in salivary glands, and it may be that sialin also makes a significant contribution to the transport of the other anions [70].

Calcium and phosphate are functionally important components of saliva, playing a crucial role in the mineralization of hard tissue and, in the case of calcium, the conformation and cross-linking of mucins [71]. Calcium appears to enter saliva predominantly as a packaged product in protein storage granules derived from the vesicular membrane-bound compartment of cells (see below). The calcium concentration of glandular saliva does not vary greatly under different stimulation conditions, and the output of calcium is well correlated with that of protein [65]. Phosphate transport by salivary glands is less well understood. There appear to have been few studies of the expression, localization and function of phosphate-transporting proteins in salivary glands. The type II sodium-phosphate cotransporter NPT2b, which transports HPO_4^{2-}, has been demonstrated in samples of human parotid and submandibular glands where it appears to be localized in basolateral membranes of acinar cells and possibly in ductal cells [72]. Type II sodium-phosphate cotransporter expression has also been characterized in goat parotid gland development [73].

Vesicular Transport and Exocytosis of Protein by Salivary Glands

Most of the protein secreted by salivary glands is derived from exocytosis of acinar cell protein storage granules [74, 75]. The fusion of storage granules with the apical membrane of acinar cells is dependent on the interaction of apical membrane SNARE proteins and secretory granule vesicle-associated membrane proteins [76]. The packaging of proteins into storage granules at high concentrations requires accumulation of charge-shielding calcium [77]. Mucins are large highly glycosylated proteins containing many posttranslational modifications including sulphation and sialylation that impart numerous negative charges. These negative charges would cause great difficulty in packaging the mucin into secretory granules prior to secretion from acinar cells. To overcome these problems, calcium ions are used to shield negative charge and fold the mucin structure into long columns [78]. Upon secretion from the acinar cell, the mucin is unfolded and becomes hydrated via the loss of calcium from the core structure [79]. Bicarbonate ions are thought to help chelate the calcium from the mucin structures [71].

It is possible to adjust conditions of parasympathetic and sympathetic stimulation so that similar quantities of protein are secreted, and under these conditions little storage granule loss is seen in acinar cells subjected to parasympathetic stimulation compared to the obvious loss following sympathetic stimulation [35]. These observations suggested a role for non-storage granule vesicular secretion in acinar cells. Vesicular transport leads to accumulation of secretory proteins in the ductal system of salivary glands in the absence of stimulation in vivo [80–82], and in vitro studies have demonstrated the rapid secretion of newly synthesized radiolabelled secretory proteins via a vesicular pathway that can be upregulated by low doses of autonomimetics [74]. The composition of proteins secreted by storage granules and vesicles differs, and the mechanisms enabling selective sequestration of different proteins are still being studied in a variety of exocrine cells including salivary acinar cells [83]. Immunogobulin A enters saliva as SIgA, a complex of dimeric IgA, J chain attached to the secretory component, the cleaved product of epithelial polymeric immunoglobulin receptor, via polymeric immunoglobulin receptor-mediated vesicular transcytosis across glandular epithelial cells. Both parasympathetic and sympathetic nerve-mediated stimuli upregulate secretion of IgA into saliva [84]. Exosomes, which originate from late endosomes and multivesicular bodies, are secreted by many different cell types and have been isolated from saliva [85, 86]. They are 30- to 100-nm cup-shaped vesicles with a lipid bilayer morphology and unlike other secretory vesicles should contain proteins like CD63 and Alix, which are characteristic of their origin from multivesicular bodies, along with genetic information of mRNA and miRNA. At present it is unclear how exosome secretion might change with physiological stimulation of salivary glands but exosomes from parotid saliva have been isolated and determined to contain a number of parotid secretory proteins in addition to proteins previously identified in exosomes from other biofluids [87].

Salivary Gland Atrophy and Regeneration

The Effects of Denervation

Studies examining how denervation alters salivary gland size and function in animal models have been reviewed previously [34]. Autologous transplantation of submandibular glands in human subjects provided an opportunity to study the effects of complete denervation. The time course of effects of denervation and re-innervation on secretion from the transplanted glands could be rationalized by comparing to earlier studies performed on animal models [88, 89]. After transplantation in humans, phases of temporal change in gland flow were observed. Firstly, secretion started almost immediately postoperatively and lasted for approximately 1 week, a pattern of secretion that was attributed to release of neurotransmitters from degenerating postganglionic axons at neuro-effector terminals. Subsequently cessation of flow occurred for several months, and this coincided with the reduction in transmitter release seen in animal studies that leads to atrophy of salivary gland secretory structures. Although atrophic, the salivary acinar cells develop a hypersensitivity to blood-borne catecholamines and the small amounts of acetylcholine spontaneously leaking from remaining postganglionic parasympathetic nerve terminals [2]. The hypersensitivity led to an epiphora, which required surgical reduction of glandular tissue, the histology of which demonstrated surviving parasympathetic ganglion cells some of which were re-innervated by sympathetic nerves, which presumably had sprouted from sites on the arteries of tissue surrounding the transplant. Previous studies of animal models have demonstrated that heterologous synaptic contacts can occur by new adrenergic axons sprouting down existing parasympathetic trunks and in time developing functional connections with parasympathetic ganglionic cells. Thus, it would appear that ganglia release chemotactic signals that lead to re-innervation.

Duct Ligation-Induced Atrophy
Salivary glands have a remarkable ability to regenerate secretory tissue following atrophy, in experimental models at least. Ligation and subsequent deligation of the main excretory duct of salivary glands is a well-studied model, which has shown functional recovery from a non-functioning state. Duct ligation-induced atrophy has been known for decades [90], and many cellular changes have been noted, in particular the rapid loss of differentiated cell types. Acini and granular ducts are no longer apparent as their secretory granules are autophagocytosed [91]. Invaginations of the plasma membrane caused by the abundant mitochondria, which are characteristic of striated ducts, are also lost. Glandular weight is reduced mainly due to loss of acinar cells following apoptosis but this is offset to some extent by proliferation of undifferentiated ductal cells [92].

Osailan et al. [93, 94] ligated rat submandibular glands from an intra-oral route without damaging the parasympathetic supply and found that even following extensive atrophy for prolonged periods the gland regenerated after removal of the ligation and eventually secreted normal amounts of saliva with a broadly normal content of ions and proteins. Regenerated glands were more responsive to low doses of cholinergic agonist indicating a possible alteration in muscarinic receptors on acinar cells. A similar approach was used by Carpenter et al. [95], and regenerated glands were found to be more responsive to parasympathetic nerve stimulation coincidental with increased density of parasympathetic nerves in the gland as demonstrated by choline acetyltransferase staining than intact, unoperated, contralateral glands suggesting that the arrangement of the parasympathetic innervation of acinar cells has been altered. One of the reasons behind the recovered secretory ability is due to the re-attachment of the parasympathetic nerves to the target cells, as shown by normal secretion in response to autonomimetic and direct nerve (parasympathetic) stimulation [95]. Recent studies have established that parasympathetic nerves and release of acetylcholine and activation of muscarinic receptors are important to the normal branching morphology of salivary glands [96] and appear to maintain an epithelial stem cell niche within salivary glands. Maintenance of glandular function is also dependent upon reciprocal signals released from epithelial cells to parasympathetic nerves; following irradiation of a developing submandibular gland, neurturin release maintains a parasympathetic innervation of effector cells [97].

The ability of the ligation/deligation model to completely regenerate is in contrast to either the partial extirpation of the gland [98] or the irradiated gland model (at higher doses) which does not recover nearly as much [99]. The lack of recovery of the irradiated gland has been attributed to loss of glandular stem cells [100]. It is interesting to note that the irradiated gland looks remarkably similar to the ligated gland (in terms of loss of acini but a proliferation of ductal cells). This suggests that the irradiated gland might go through a similar atrophic process as the ligated gland. Certainly ligation-induced atrophy is faster acting than a disuse atrophy seen when the parasympathetic nerves are cut even though some autophagic loss of aquaporin 5 and other proteins occurs [101]. Perhaps the reason that partial extirpation of glands does not cause much glandular regeneration is because the damaged gland does not go through an atrophic process. Our recent studies have shown that the atrophic process is associated with the activation of the mTOR (mammalian targets of rapamycin) pathway [92]. This was a surprise since mTOR is normally associated with cell and tumour growth [102]. In addition, autophagy, the self-ingestion of secretory granules, is well documented to inhibit mTOR [103] yet in our study the two processes appeared in acinar cells at the same time. This apparent disparity may have been resolved by the demonstration of these potentially antagonistic mechanisms in spatially separated parts of the same cell [103].

Conclusions

The peripheral control of salivary gland function is the result of communication between autonomic nerves and the parenchymal and vascular compartments with which they are intimately associated. Oral dysfunction resulting from chronic loss of saliva is most frequently associated with pharmacological disruption of nerve signalling. Peripheral signalling is dominated by the cholinergic mechanism, and anticholinergic drugs can cause profound oral dryness. Most drugs causing salivary hypofunction act centrally but we know less about the mechanisms of interruption and clearly we need to increase our knowledge of the central connections regulating secretion by salivary glands in order to be able to address the problem of dry mouth. Salivary gland inflammatory and irradiation-induced disease is associated with atrophy. Increased understanding of the trophic influences of autonomic nerve signalling should benefit the development of therapeutic approaches to glandular regeneration.

References

1 Phillips CJ, Weiss A, Tandler B: Plasticity and patterns of evolution in mammalian salivary glands: comparative immunohistochemistry of lysozyme in bats. Eur J Morphol 1998;36:19–26.

2 Emmelin N: Nerve interactions in salivary-glands. J Dent Res 1987;66:509–517.

3 Dawes C: Circadian rhythms in human salivary flow rate and composition. J Physiol 1972;220:529–545.

4 Hector MP, Garrett JR, Ekstrom J, Anderson LC: Reflexes of salivary secretion; in Garrett JR, Ekstrom J, Anderson LC (eds): Neural Mechanisms of Salivary Gland Secretion. Basel, Karger, 1999, pp 196–217.

5 Speirs RL: Secretion of saliva by human lip mucous glands and parotid glands in response to gustatory stimuli and chewing. Arch Oral Biol 1984;29:945–948.

6 Boros I, Keszler P, Zelles T: Study of saliva secretion and the salivary fluoride concentration of the human minor labial glands by a new method. Arch Oral Biol 1999;44(suppl 1):S59–S62.

7 Veerman ECI, van den Keybus PAM, Vissink A, Amerongen AVN: Human glandular salivas: their separate collection and analysis. Eur J Oral Sci 1996; 104:346–352.

8 Lee VM, Linden RWA: An olfactory submandibular salivary reflex in humans. Exp Physiol 1992;77:221–224.

9 Dawes C, O'Connor AM, Aspen JM: The effect on human salivary flow rate of the temperature of a gustatory stimulus. Arch Oral Biol 2000;45:957–961.

10 Wang B, Danjo A, Kajiya H, Okabe K, Kido MA: Oral epithelial cells are activated via TRP channels. J Dent Res 2011;90:163–167.

11 Lorenz K, Bader M, Klaus A, Weiss W, Gorg A, Hofmann T: Orosensory stimulation effects on human saliva proteome. J Agr Food Chem 2011;59:10219–10231.

12 Dunerengstrom M, Fredholm BB, Larsson O, Lundberg JM, Saria A: Autonomic mechanisms underlying capsaicin induced oral sensations and salivation in man. J Physiol (London) 1986;373:87–96.

13 Bradley RM, Fukami H, Suwabe T: Neurobiology of the gustatory-salivary reflex. Chem Senses 2005;30:I70–I71.

14 Matsuo R: Central connections for salivary innervations and efferent impulse formation; in Garrett JR, Ekstrom J, Anderson LC (eds): Neural Mechanisms of Salivary Gland Secretion. Basel, Karger, 1999, pp 26–43.

15 Khosravani N, Sandberg M, Ekstrom J: The otic ganglion in rats and its parotid connection: cholinergic pathways, reflex secretion and a secretory role for the facial nerve. Exp Physiol 2006;91:239–247.

16 Ishizuka KI, Oskutyte D, Satoh Y, Murakami T: Multi-source inputs converge on the superior salivatory nucleus neurons in anaesthetized rats. Auton Neurosci Basic Clin 2010;156:104–110.

17 Ueda H, Mitoh Y, Fujita M, et al: Muscarinic receptor immunoreactivity in the superior salivatory nucleus neurons innervating the salivary glands of the rat. Neurosci Lett 2011;499:42–46.

18 Renzi A, De Luca LA Jr, Menani JV: Lesions of the lateral hypothalamus impair pilocarpine-induced salivation in rats. Brain Res Bull 2002;58:455–459.

19 Takakura AC, Moreira TS, De Luca LA Jr, Renzi A, Menani JV, Colombari E: Effects of AV3V lesion on pilocarpine-induced pressor response and salivary gland vasodilation. Brain Res 2005;1055:111–121.

20 Kringelbach ML, O'Doherty J, Rolls ET, Andrews C: Activation of the human orbitofrontal cortex to a liquid food stimulus is correlated with its subjective pleasantness. Cereb Cortex 2003;13:1064–1071.

21 Rolls ET: Taste, olfactory and food texture reward processing in the brain and obesity. Int J Obes 2011;35:550–561.

22 Spence C: Mouth-watering: the influence of environmental and cognitive factors on salivation and gustatory/flavor perception. J Texture Stud 2011;42:157–171.

23 Small DM: Taste representation in the human insula. Brain Struct Funct 2010;214:551–561.

24 Ilangakoon Y, Carpenter GH: Is the mouthwatering sensation a true salivary reflex? J Texture Stud 2011;42:212–216.

25 Moreira TS, Takakura AC, Colombari E, De Luca LA Jr, Renzi A, Menani JV: Central moxonidine on salivary gland blood flow and cardiovascular responses to pilocarpine. Brain Res 2003;987:155–163.

26 Phillips MA, Szabadi E, Bradshaw CM: Comparison of the effects of clonidine and yohimbine on pupillary diameter at different illumination levels. Br J Clin Pharmacol 2000;50:65–68.

27 Gotrick B, Giglio D, Tobin G: Effects of amphetamine on salivary secretion. Eur J Oral Sci 2009;117:218–223.

28 Garrett JR: The proper role of nerves in salivary secretion – a review. J Dent Res 1987;66:387–397.

29 Garrett JR, Kidd A: The innervation of salivary-glands as revealed by morphological methods. Microsc Res Tech 1993;26:75–91.

30 Garrett JR, Anderson LC: Rat sublingual salivary-glands – secretory changes on parasympathetic or sympathetic-nerve stimulation and a reappraisal of the adrenergic-innervation of striated ducts. Arch Oral Biol 1991;36:675–683.

31 Rossoni RB, Machado AB, Machado CRS: Histochemical-study of catecholamines and cholinesterases in the autonomic nerves of the human minor salivary-glands. Histochem J 1979;11:661–668.

32 Ekstrom J: Role of nonadrenergic, noncholinergic autonomic transmitters in salivary glandular activities in vivo; in Garrett JR, Ekstrom J, Anderson LC (eds): Neural Mechanisms of Salivary Gland Secretion. Basel, Karger, 1999, pp 94–130.

33 Kusakabe T, Matsuda H, Gono Y, et al: Distribution of VIP receptors in the human submandibular gland: an immunohistochemical study. Histol Histopathol 1998;13:373–378.

34 Proctor GB, Carpenter GH: Regulation of salivary gland function by autonomic nerves. Auton Neurosci 2007;133:3–18.

35 Asking B, Gjorstrup P: Synthesis and secretion of amylase in the rat parotid gland following autonomic nerve stimulation in vivo. Acta Physiol Scand 1987;130:439–445.

36 Anderson LC, Garrett JR, Zhang X, Proctor GB, Shori DK: Differential secretion of proteins by rat submandibular acini and granular ducts on graded autonomic nerve stimulations. J Physiol 1995;485:503–511.

37 Carpenter GH, Proctor GB, Anderson LC, Zhang XS, Garrett JR: Immunoglobulin A secretion into saliva during dual sympathetic and parasympathetic nerve stimulation of rat submandibular glands. Exp Physiol 2000;85:281–286.

38 Matsuo R, Garrett JR, Proctor GB, Carpenter GH: Reflex secretion of proteins into submandibular saliva in conscious rats, before and after preganglionic sympathectomy. J Physiol 2000;527:175–184.

39 Culp DJ, Graham LA, Latchney LR, Hand AR: Rat sublingual gland as a model to study glandular mucous cell secretion. Am J Physiol 1991;260:C1233–C1244.

40 Baum BJ, Wellner RB: Receptors in salivary glands; in Garrett JR, Ekstrom J, Anderson LC (eds): Neural Mechanisms of Salivary Gland Secretion. Basel, Karger, 1999, pp 44–58.

41 Nakamura T, Matsui M, Uchida K, et al: M_3 muscarinic acetylcholine receptor plays a critical role in parasympathetic control of salivation in mice. J Physiol 2004;558:561–575.

42 Gautam D, Heard TS, Cui Y, Miller G, Bloodworth L, Wess J: Cholinergic stimulation of salivary secretion studied with M_1 and M_3 muscarinic receptor single- and double-knockout mice. Mol Pharmacol 2004;66:260–267.

43 Gallacher DV, Smith PM: Autonomic transmitters and Ca^{2+}-activated cellular responses to salivary glands in vitro; in Garrett JR, Ekstrom J, Anderson LC (eds): Neural Mechanisms of Salivary Gland Secretion. Basel, Karger, 1999, pp 80–93.

44 Ambudkar IS: Polarization of calcium signaling and fluid secretion in salivary gland cells. Curr Med Chem 2012;19:5774–5781.

45 Melvin JE, Yule D, Shuttleworth T, Begenisich T: Regulation of fluid and electrolyte secretion in salivary gland acinar cells. Annu Rev Physiol 2005;67:445–469.

46 Huang GN, Zeng W, Kim JY, et al: STIM1 carboxyl-terminus activates native SOC, I(crac) and TRPC1 channels. Nat Cell Biol 2006;8:1003–1010.

47 Ong HL, Cheng KT, Liu X, et al: Dynamic assembly of TRPC1-STIM1-Orai1 ternary complex is involved in store-operated calcium influx. Evidence for similarities in store-operated and calcium release-activated calcium channel components. J Biol Chem 2007;282:9105–9116.

48 Pani B, Ong HL, Brazer SC, et al: Activation of TRPC1 by STIM1 in ER-PM microdomains involves release of the channel from its scaffold caveolin-1. Proc Natl Acad Sci USA 2009;106:20087–20092.

49 Moller K, Benz D, Perrin D, Soling HD: The role of protein kinase C in carbachol-induced and of cAMP-dependent protein kinase in isoproterenol-induced secretion in primary cultured guinea pig parotid acinar cells. Biochem J 1996 15; 314:181–187.

50 Asking B: Sympathetic stimulation of amylase secretion during a parasympathetic background activity in the rat parotid gland. Acta Physiol Scand 1985; 124:535–542.

51 Tanimura A, Nezu A, Tojyo Y, Matsumoto Y: Isoproterenol potentiates alpha-adrenergic and muscarinic receptor-mediated Ca^{2+} response in rat parotid cells. Am J Physiol 1999;276:C1282– C1287.

52 Bobyock E, Chernick WS: Vasoactive intestinal peptide interacts with alpha-adrenergic-, cholinergic-, and substance-P-mediated responses in rat parotid and submandibular glands. J Dent Res 1989;68:1489–1494.

53 Straub SV, Giovannucci DR, Bruce JI, Yule DI: A role for phosphorylation of inositol 1,4,5-trisphosphate receptors in defining calcium signals induced by peptide agonists in pancreatic acinar cells. J Biol Chem 2002;277:31949–31956.

54 Proctor GB, Asking B: A comparison between changes in rat parotid protein-composition 1 and 12 weeks following surgical sympathectomy. Q J Exp Physiol 1989;74:835–840.

55 Carpenter GH, Proctor GB, Garrett JR: Preganglionic parasympathectomy decreases salivary SIgA secretion rates from the rat submandibular gland. J Neuroimmunol 2005;160:4–11.

56 Baker OJ: Tight junctions in salivary epithelium. J Biomed Biotechnol 2010; 2010:278948.

57 Bundgaard M, Moller M, Poulsen JH: Localization of sodium pump sites in cat salivary glands. J Physiol 1977;273:339–353.

58 Roussa E: Channels and transporters in salivary glands. Cell Tissue Res 2011; 343:263–287.

59 Martinez JR, Holzgreve H, Frick A: Micropuncture study of submaxillary glands of adult rats. Pflugers Arch Gesamte Physiol Menschen Tiere 1966;290: 124–133.

60 Romanenko VG, Catalan MA, Brown DA, et al: Tmem16A encodes the Ca^{2+}-activated Cl^- channel in mouse submandibular salivary gland acinar cells. J Biol Chem 2010;285:12990–13001.

61 Evans RL, Turner RJ: New insights into the upregulation and function of the salivary Na^+-K^+-$2Cl^-$ cotransporter. Eur J Morphol 1998;36(suppl):142–146.

62 Gresz V, Kwon TH, Hurley PT, et al: Identification and localization of aquaporin water channels in human salivary glands. Am J Physiol Gastrointest Liver Physiol 2001;281:G247–G254.

63 Ma T, Song Y, Gillespie A, Carlson EJ, Epstein CJ, Verkman AS: Defective secretion of saliva in transgenic mice lacking aquaporin-5 water channels. J Biol Chem 1999;274:20071–20074.

64 Thaysen JH, Thorn NA, Schwartz IL: Excretion of sodium, potassium, chloride and carbon dioxide in human parotid saliva. Am J Physiol 1954;178:155–159.

65 Young JA, Schneyer CA: Composition of saliva in mammalia. Aust J Exp Biol Med Sci 1981;59:1–53.

66 Winston DC, Schulte BA, Garrett JR, Proctor GB: Na^+,K^+-ATPase in cat salivary glands and changes induced by nerve stimulation: an immunohistochemical study. J Histochem Cytochem 1990;38:1187–1191.

67 Lee MG, Ohana E, Park HW, Yang D, Muallem S: Molecular mechanism of pancreatic and salivary gland fluid and HCO_3 secretion. Physiol Rev 2012;92: 39–74.

68 Aboud Z, Misra S, Warner T, et al: The enterosalivary bioconversion of nitrate to nitrite underlies the blood pressure (BP) lowering and anti-platelet effects of a dietary nitrate load. Br J Clin Pharmacol 2008;65:999.

69 Stephen KW, Robertson JW, Harden RM, Chisholm DM: Concentration of iodide, pertechnetate thiocyanate, and bromide in saliva from parotid, submandibular, and minor salivary glands in man. J Lab Clin Med 1973;81:219–229.

70 Qin L, Liu X, Sun Q, et al: Sialin (SL-C17A5) functions as a nitrate transporter in the plasma membrane. Proc Natl Acad Sci USA 2012;109:13434–13439.

71 Quinton PM: Role of epithelial HCO_3^- transport in mucin secretion: lessons from cystic fibrosis. Am J Physiol Cell Physiol 2010;299:C1222–C1233.

72 Homann V, Rosin-Steiner S, Stratmann T, Arnold WH, Gaengler P, Kinne RK: Sodium-phosphate cotransporter in human salivary glands: molecular evidence for the involvement of NPT2b in acinar phosphate secretion and ductal phosphate reabsorption. Arch Oral Biol 2005;50:759–768.

73 Huber K, Roesler U, Muscher A, et al: Ontogenesis of epithelial phosphate transport systems in goats. Am J Physiol Regul Integr Comp Physiol 2003;284: R413–R421.

74 Huang AY, Castle AM, Hinton BT, Castle JD: Resting (basal) secretion of proteins is provided by the minor regulated and constitutive-like pathways and not granule exocytosis in parotid acinar cells. J Biol Chem 2001;276:22296–22306.

75 Segawa A, Loffredo F, Puxeddu R, Yamashina S, Testa Riva F, Riva A: Cell biology of human salivary secretion. Eur J Morphol 2000;38:237–241.

76 Turner RJ, Sugiya H: Understanding salivary fluid and protein secretion. Oral Dis 2002;8:3–11.

77 Verdugo P: Mucin exocytosis. Am Rev Respir Dis 1991;144:S33– S37.

78 Ambort D, Johansson MEV, Gustafsson JK, et al: Calcium and pH-dependent packing and release of the gel-forming MUC2 mucin. Proc Natl Acad Sci USA 2012;109:5645–5650.

79 Kesimer M, Makhov AM, Griffith JD, Verdugo P, Sheehan JK: Unpacking a gel-forming mucin: a view of MUC5B organization after granular release. Am J Physiol Lung Cell Mol Physiol 2010; 298:L15–L22.

80 Garrett JR, Zhang XS, Proctor GB, Anderson LC, Shori DK: Apical secretion of rat submandibular tissue kallikrein continues in the absence of external stimulation: evidence for a constitutive secretory pathway. Acta Physiol Scand 1996; 156:109–114.

81 Garrett JR, Suleiman AM, Anderson LC, Proctor GB: Secretory responses in granular ducts and acini of submandibular glands in vivo to parasympathetic or sympathetic nerve stimulation in rats. Cell Tissue Res 1991;264:117–126.

82 Proctor GB, Carpenter GH, Segawa A, Garrett JR, Ebersole L: Constitutive secretion of immunoglobulin A and other proteins into lumina of unstimulated submandibular glands in anaesthetised rats. Exp Physiol 2003;88:7–12.

83 Gorr SU, Venkatesh SG, Darling DS: Parotid secretory granules: crossroads of secretory pathways and protein storage. J Dent Res 2005;84:500–509.

84 Proctor GB, Carpenter GH: Neural control of salivary S-IgA secretion. Int Rev Neurobiol 2002;52:187–212.

85 Berckmans RJ, Sturk A, van Tienen LM, Schaap MC, Nieuwland R: Cell-derived vesicles exposing coagulant tissue factor in saliva. Blood 2011;117:3172–3180.

86 Palanisamy V, Sharma S, Deshpande A, Zhou H, Gimzewski J, Wong DT: Nanostructural and transcriptomic analyses of human saliva derived exosome. Plos One 2010;5:e8577.

87 Gonzalez-Begne M, Lu B, Han X, et al: Proteomic analysis of human parotid gland exosomes by multidimensional protein identification technology (MudPIT). J Proteome Res 2009;8:1304–1314.

88 Geerling G, Garrett JR, Paterson KL, et al: Innervation and secretory function of transplanted human submandibular salivary glands. Transplantation 2008; 85:135–140.

89 Borrelli M, Schroder C, Dart JK, et al: Long-term follow-up after submandibular gland transplantation in severe dry eyes secondary to cicatrizing conjunctivitis. Am J Ophthalmol 2010;150:894–904.

90 Ohlin P, Perec C: Secretory responses and choline acetylase of the rat's submaxillary gland after duct ligation. Experientia 1967;23:248–249.

91 Silver N, Proctor GB, Arno M, Carpenter GH: Activation of mTOR coincides with autophagy during ligation-induced atrophy in the rat submandibular gland. Cell Death Dis 2010;1:e14.

92 Takahashi S, Nakamura S, Suzuki R, et al: Apoptosis and mitosis of parenchymal cells in the duct-ligated rat submandibular gland. Tissue Cell 2000;32:457–463.

93 Osailan SM, Proctor GB, McGurk M, Paterson KL: Intraoral duct ligation without inclusion of the parasympathetic nerve supply induces rat submandibular gland atrophy. Int J Exp Pathol 2006;87:41–48.

94 Osailan SM, Proctor GB, Carpenter GH, Paterson KL, McGurk M: Recovery of rat submandibular salivary gland function following removal of obstruction: a sialometrical and sialochemical study. Int J Exp Pathol 2006;87:411–423.

95 Carpenter GH, Khosravani N, Ekstrom J, Osailan SM, Paterson KP, Proctor GB: Altered plasticity of the parasympathetic innervation in the recovering rat submandibular gland following extensive atrophy. Exp Physiol 2009;94:213–219.

96 Knox SM, Lombaert IMA, Reed X, Vitale-Cross L, Gutkind JS, Hoffman MP: Parasympathetic innervation maintains epithelial progenitor cells during salivary organogenesis. Science 2010;329: 1645–1647.

97 Knox SM, Lombaert IMA, Haddox CL, et al: Parasympathetic stimulation improves epithelial organ regeneration. Nat Commun 2013;4:1494.

98 Takahashi S, Wakita M: Regeneration of the intralobular duct and acinus in rat submandibular glands after YAG laser irradiation. Arch Histol Cytol 1993;56:199–206.

99 Konings AWT, Coppes RP, Vissink A: On the mechanism of salivary gland radiosensitivity. Int J Radiat Oncol 2005;62:1187–1194.

100 Nanduri LSY, Maimets M, Pringle SA, van der Zwaag M, van Os RP, Coppes RP: Regeneration of irradiated salivary glands with stem cell marker expressing cells. Radiother Oncol 2011;99:367–372.

101 Azlina A, Javkhlan P, Hiroshima Y, et al: Roles of lysosomal proteolytic systems in AQP5 degradation in the submandibular gland of rats following chorda tympani parasympathetic denervation. Am J Physiol Gastrointest Liver Physiol 2010;299:G1106–G1117.

102 Wang XM, Proud CG: The mTOR pathway in the control of protein synthesis. Physiology 2006;21:362–369.

103 Diaz-Troya S, Perez-Perez ME, Florencio FJ, Crespo JL: The role of TOR in autophagy regulation from yeast to plants and mammals. Autophagy 2008; 4:851–865.

104 Kalk WW, Vissink A, Spijkervet FK, Bootsma H, Kallenberg CG, Nieuw Amerongen AV: Sialometry and sialochemistry: diagnostic tools for Sjogren's syndrome. Ann Rheum Dis 2001;60:1110–1116.

Gordon Proctor
Salivary Research Unit, Floor 17 Tower Wing
King's College London Dental Institute, Guy's and St Thomas' Hospitals
Great Maze Pond, London SE1 9RT (UK)
E-Mail gordon.proctor@kcl.ac.uk

Ligtenberg AJM, Veerman ECI (eds): Saliva: Secretion and Functions.
Monogr Oral Sci. Basel, Karger, 2014, vol 24, pp 30–39 (DOI: 10.1159/000358782)

Salivary Pellicles

Liselott Lindh[a] · Watcharapong Aroonsang[a, c] · Javier Sotres[b] ·
Thomas Arnebrant[b]

[a]Department of Prosthetic Dentistry, Faculty of Odontology, and [b]Department of Biomedical Sciences,
Faculty of Health and Society, Malmö University, Malmö, Sweden; [c]Department of Prosthodontics,
Faculty of Dentistry, Chiang Mai University, Chiang Mai, Thailand

Abstract

The salivary pellicle is a thin acellular organic film that forms on any type of surface upon exposure to saliva. The role of the pellicle is manifold, and it plays an important role in the maintenance of oral health. Its functions include not only substratum protection and lubrication, but also remineralization and hydration. It also functions as a diffusion barrier and possesses buffering ability. Not only the function, but also the formation, composition and stability of the pellicle are known to be highly influenced by the physicochemical properties of both substrata and ambient media. In this chapter, we discuss these aspects of salivary pellicles, an area where research has boomed in the past years partly because of the application of experimental techniques often reserved for more traditional surface science studies.

The salivary pellicle, a thin acellular film predominantly consisting of salivary proteins, covers surfaces that are exposed in the oral cavity. It is known to have an important role in maintaining oral health [1, 2]. The salivary pellicle was first described by Nasmyth in 1839 [3]. It was thought to emerge from embryonic origin for a long time until Dawes et al. [4], in 1963, clearly clarified that organic structures cover the teeth into the embryonic-origin membranes and acquired membranes after tooth eruption. Since then, the latter membrane is called the acquired enamel pellicle.

Surfaces in the oral cavity, i.e. natural surfaces such as enamel, dentine and oral mucosa, as well as artificial ones such as dental materials, are all covered by salivary pellicles. Therefore, studying the interactions between salivary proteins and oral surfaces is the key for understanding the formation of salivary pellicles. This knowledge will also be useful to comprehend the complexity of oral pathological conditions and lead to, for instance, the development of new dental materials, material coatings and more effective dental care products and saliva substitutes.

This chapter discusses salivary pellicle characteristics such as formation, structure, composition as well as chemical and mechanical stability from a physicochemical point of view.

The Salivary Pellicle Formation

The formation of salivary pellicles is a highly selective adsorption process where macromolecules from whole saliva adsorb onto oral surfaces [5, 6]. The initial adsorption of the salivary pellicle takes place within seconds of exposure to whole saliva [2, 7, 8]. The pellicle reaches a thickness of 10–20 nm within a couple of minutes [9]. Phosphoproteins, such as acidic proline-rich proteins, statherins and histatins, have been shown to be among the initially adsorbed proteins onto hydroxyapatite (HA) surfaces in in vitro studies [10, 11]. However, in vivo studies revealed a more diverse and complex pellicle composition, with the presence of high-molecular-weight glycoproteins (MUC5B and MUC7), amylase, cystatins, lysozyme and lactoferrin [2]. It is accepted that electrostatic interactions play an important role in the formation of the initial pellicle on enamel surfaces [10]; however, the widely different net charges of the components in the film strongly indicate contributions also by other interactions such as van der Waals and hydrophobic interactions [8, 12].

The initial adsorption is followed by a second stage or 'maturation' where protein aggregates are formed via protein-protein interactions [2, 13]. Then, the pellicle thickness may increase up to 100 or 1,000 nm, depending on the location within the mouth [14]. The formation of protein aggregates, which could be heterotypic complexes [15] and/or micelle-like structures [16], has previously been reported and might be an explanation for the reported increased thickness [2]. These protein aggregates are also described as 'supramolecular precursors' [17].

The formation of salivary pellicles in the oral cavity is a dynamic process, which is continuously altered by many factors, for instance, heterotypic complexing, enzymatic cross-linking, proteolytic degradation as well as the modification of adsorbed molecules by components originating from oral bacteria, as well as salivary flow and clearance [9]. Despite the fact that more clinically accurate information of the pellicle could be obtained from in vivo studies, in vitro studies are useful to obtain more fundamental information regarding pellicle characteristics with the aim to enhance the knowledge in the in vivo situation. Several surface models have been employed in in vitro studies to investigate the adsorption behaviour in order to gain improved understanding of the saliva-material interactions. Studies of saliva adsorption on germanium prisms [7], hydrophilic and hydrophobized silica surfaces [8, 12, 18–23], titanium surfaces [23, 24] and additionally also on gold and zirconia have been performed [23].

Research regarding saliva, pellicles and oral interfaces with a physicochemical approach has been going on for more than 40 years at Malmö University [7–8, 12, 18–23, 25–29, 36–37]. In two key studies [12, 19] we showed, by means of in situ ellipsometry experiments, differences in adsorption kinetics and total adsorbed amounts of saliva on surfaces with different wettability, i.e. hydrophilic and hydrophobized silica surfaces. An example of such a measurement is shown in figure 1a. The adsorption on the hydrophilic surface was initially rapid, followed by a slow but continuous increase in the build-up of the adsorbed layer, whereas on the hydrophobic substratum a defined plateau was reached more quickly [19]. A higher amount of material was adsorbed on hydrophobized than on hydrophilic surfaces [8, 19]. In general, driving forces for protein adsorption are considered to be mainly electrostatic and hydrophobic interactions, sometimes in combination with effects by the conformational entropy gain upon unfolding [30]. Both silica and hydrophobized silica have comparable surface potentials [31]. Therefore, the higher adsorbed amount on the hydrophobized surfaces indicates the importance of hydrophobic interactions. In a separate study [29], we also showed that the adsorption from human whole saliva as well as effects on the adsorbed layer by addition of sodium dodecyl sulphate (SDS) was different

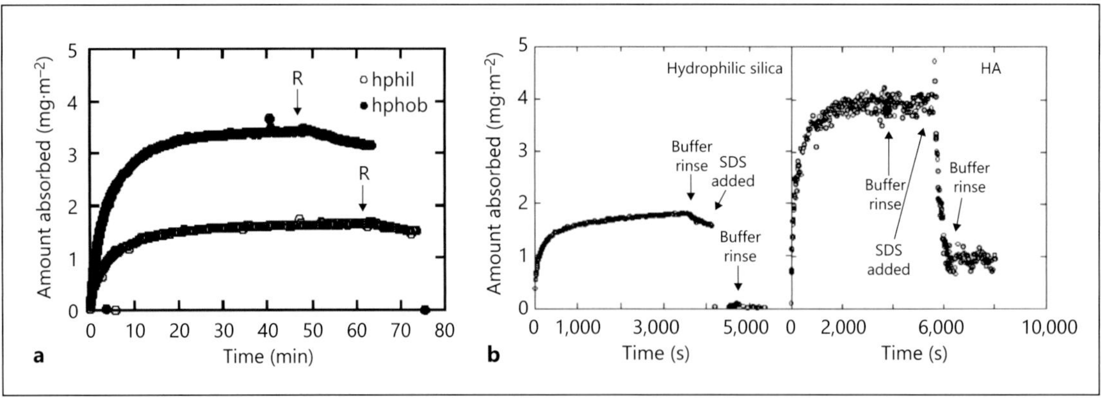

Fig. 1. Adsorbed amounts of human whole-saliva pellicle versus time measured by in situ ellipsometry on (**a**), hydrophilic (○) and hydrophobized (●) silica surfaces where R indicates buffer rinse (reprinted with permission from Lindh et al. [19], copyright 1999 Taylor & Francis) and (**b**), hydrophilic silica and HA surfaces including also buffer rinse, addition of sodium dodecyl sulphate (SDS) and a final step of buffer rinse (reprinted with permission from Santos et al. [29], copyright 2010 Taylor & Francis).

on hydrophilic HA surfaces compared to that on hydrophilic silica surfaces (fig. 1b). This clearly shows that not only wettability, but also chemical composition and hence surface charge density influence the salivary adsorption [8, 12, 19, 29].

Salivary Pellicle Structure

Various approaches have been employed to characterize the morphology and ultrastructure of salivary pellicles, including transmission and scanning electron microscopy, field emission in-lens scanning electron microscopy, cryo-electron microscopy, surface force measurements and neutron reflection [14, 26, 32–36]. Recently, atomic force microscopy has provided valuable information of the pellicle topography as well [20, 26].

Even though there seems to be variation of the findings on the ultrastructure of salivary pellicles due to methodological differences, the structure of salivary pellicles has been generally described as a dense inner layer and a loosely arranged outer layer [8, 20, 26]. This ultrastructural finding is supported by previous observations from in vivo and in vitro studies. For in-

stance, a transmission electron microscopy study of in vivo salivary films formed on bovine enamel indicated that the pellicle is composed of an outer globular and an inner dense basal layer [33]. An in vitro study by Cardenas et al. [26], using neutron reflectivity, ellipsometry and atomic force microscopy, showed that the morphology and density profile of salivary films formed on model alumina surfaces, which have a similar iso-electric point as enamel, could be described as two layers, i.e. an inner dense and thin uniform layer of around 4 nm, and an outer diffuse and thick layer of approximately 30 nm with projections of macromolecules into the bulk solution. Another study by Santos et al. [29] also supports the biphasic structure of salivary films on HA and silica surfaces. In this study the results indicated that the inner layer, which was strongly attached to the surface, contained a low amount of water in accordance with what is expected for a dense monolayer of proteins. The outer layer mainly consisted of more flexible and loosely bound proteins, and had a larger amount of trapped water. According to the findings in these studies, the thin layer forms fast, which is in line with the general adsorption behaviour of polydisperse polymers, suggesting that smaller, fast diffusing molecules, like acidic proline-rich proteins and statherins, instantly adsorb to the solid surface. Later on, large, slowly diffusing molecules with high surface affinity, like mucins, would replace the adsorbed smaller molecules [37], in line with what has been described for blood proteins, the so-called Vroman effect [38]. In the same study by Santos et al. [29], the thickness of salivary films was determined. The results were in good agreement with other studies [8, 20, 26, 36] (approx. 30 nm) where the thickness of films formed on silica, mica and alumina surfaces was studied. Moreover, these in vitro findings are in accordance with the in vivo pellicle thickness that has been reported to range between 30 and 100 nm [9].

A factor that has to be considered when modelling the structure of the initial adsorbed pellicle is that it has to be compatible with enamel homeostasis. The thermodynamically more stable phase of calcium phosphate, HA, is the main building block in the outer layer of teeth, i.e. the enamel [39]. Enamel is constantly subjected to dynamic mineralization/demineralization processes [40]. On the one hand, enamel demineralizes under acidic conditions, which can be induced by the dental plaque upon the consumption of sugars [41] or by the intake of soft drinks [42]. On the other hand, bulk saliva is supersaturated with respect to calcium and phosphate ions. However, nature provides an optimal homeostatic control mechanism by saliva. Besides its buffering properties [43], saliva also slows the acid-induced dissolution rate of enamel through the formation of the pellicle at its surface [44]. Moreover, the pellicle prevents the continuous precipitation of calcium phosphate salts onto the enamel surface [45]. Interestingly, pellicle components are also known to stabilize calcium phosphate clusters in bulk saliva [46, 47]. Along with these stabilizing properties, the pellicle allows selective ion exchange at the enamel surface. While the mechanism by which this occurs is still unclear, it has

been proposed [48] that the pellicle is in reality an open structure that would allow such an exchange.

Salivary Pellicle Composition

Several studies have been performed to identify the composition of salivary pellicles. Here the limitations of in vivo studies are the minute quantities of collected pellicle material and the difficulties in harvesting techniques that are harmless to the oral surfaces [2]. To overcome these problems, studies using enamel slabs mounted in intra-oral appliances allow pellicles to be collected from many individuals, and by several collection periods [49]. Additionally, a validation of the harvesting technique can be completed without any damage on the individual's own teeth [27]. Although in vitro studies are not able to perfectly mimic the dynamic conditions in the oral cavity, these studies are valuable for a more fundamental understanding of salivary pellicles. A main advantage is that the day-to-day and/or interindividual differences in pellicle compositions can be eliminated.

Various pellicle-harvesting techniques have been introduced such as for instance the combination of mechanical and chemical collecting methods [27, 50, 51]. The compositions of the harvested pellicles have been determined by a combination of various analytical approaches, including electrophoresis, histological staining, immunogenic and chromatographic techniques. Results suggest that for example amylase, albumin, immunoglobulins, the glycoprotein gp340 (salivary agglutinin), proline-rich proteins, statherins, histatin 1, MUC5B, carbonic anhydrases, lactoferrin, lysozyme and cystatins are included in the in vivo pellicle [13]. In figure 2, two-dimensional SDS polyacrylamide gel electrophoresis (2-D SDS-PAGE) from both whole saliva (for comparison) and in vivo pellicles is shown. The abundance of proteins and the complexity of

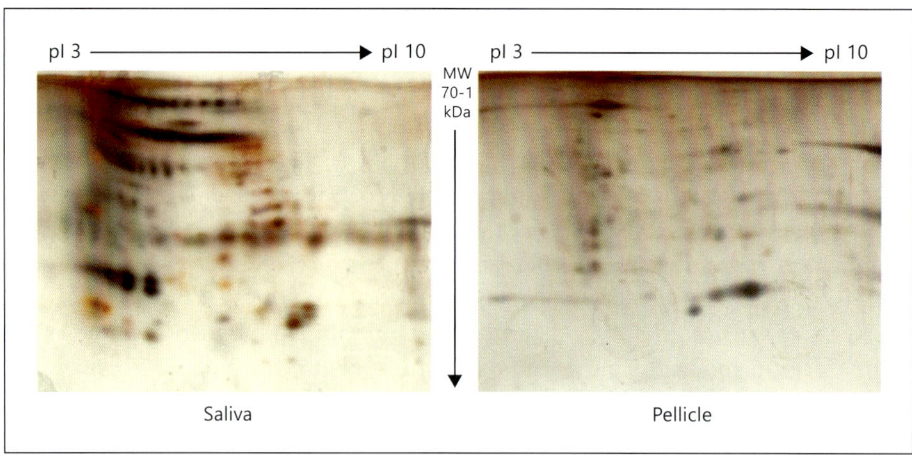

Fig. 2. Coomassie brilliant blue- and silver-stained 2-D SDS-PAGE from whole saliva and an in vivo pellicle collected by mechanically assisted SDS elution. For the first dimension, iso-electric points of proteins were separated in the horizontal direction (from pI 3 to pI 10), and for the second dimension, the sizes of the proteins were separated in the vertical direction (with the given size of the standard in kilodaltons on the right) (the pellicle picture reprinted with permission from Svendsen et al. [27], copyright 2008 Taylor & Francis).

pellicle compositions harvested by means of mechanically assisted SDS elution are evident [27]. Recently, novel proteomic techniques, including liquid chromatography-electrospray ionization tandem mass spectrometry and matrix-assisted laser desorption ionization time-of-flight mass spectrometry, have facilitated a more detailed identification of the pellicle composition; they are summarized in Siqueira et al. [13].

The composition of the pellicle is influenced by several factors including the properties of its substratum. A study by Svendsen and Lindh [28] indicated that surface free energy, surface chemical composition and surface roughness play important roles in the composition of the salivary pellicles. In this study, pellicles formed on human enamel, titanium and (poly)methylmethacrylate were harvested and 2-D SDS-PAGE used to elucidate the differences in their protein patterns. The results from Svendsen and Lindh [28] indicated some discrepancies in the compositions among the surfaces. When comparing with identified proteins in 2-D SDS-PAGE protein maps of human saliva from previous reports [52, 53], it was noted that for example on human enamel the major adsorbed proteins are likely carbonic anhydrase VI, cystatin SN, lysozyme and histatins, whereas on (poly)methylmethacrylate the main components are likely lysozyme and histatin.

These findings clearly indicate the influence of the underlying surface characteristics, such as surface free energy and charge, on the differences in pellicle composition and film build-up. Although a higher surface roughness has been shown to have an association with increased protein adsorption [54], the surface roughnesses of (poly)methylmethacrylate and titanium used in this study [28] were similar, indicating that the compositional difference of the pellicles between these two surfaces was possibly due to the chemistry of substrata. It is likely that surface roughness may be influential, but it is not the only determinant for pellicle composition.

Even though the major components of salivary pellicles are proteins, also carbohydrates and lipids are found in the pellicle [2]. The lipids in enamel pellicles have been shown to have an acid protection ability on the enamel surface suggested to be caused by retarding the lactic acid diffusion [55]. However, studies regarding the lipids present in the salivary pellicles are scarce [56], and further research in this field is needed for deeper insight into the roles of lipids within the pellicle.

Chemical and Mechanical Stability of Salivary Pellicles

An important property of the pellicle is its resistance to both chemical and mechanical attacks. This is relevant as, under in vivo conditions, the pellicle is subjected to these two types of challenges. Furthermore, knowledge on this resistance is important as it provides an understanding of pellicle elution procedures. The latter may be induced in vivo by surfactants present in oral care products, and are also a prerequisite for surfactant elution of pellicles in compositional analysis.

Numerous investigations have been performed addressing interactions between surfactants and salivary pellicles formed from whole and glandular secretions [8, 18, 23, 29]. The overall conclusion is that the major parts of the pellicles are detached and that the elutable fractions depend on the type of surfactant as well as the surface type. Santos et al. compared the effect of two surfactants used in oral hygiene products, SDS and delmopinol, on pellicles formed on HA and silica surfaces [29]. They found that SDS was more efficient than delmopinol in removing the salivary film from both silica and HA surfaces. The findings also revealed that both surfactants could remove more of the adsorbed film on silica surfaces compared to on HA surfaces. On the other hand, when comparing SDS elutability of salivary films formed on HA with zirconia and titanium surfaces, respectively, HA was suggested to be the most easily cleaned surface [23].

Oral surfaces are constantly exposed to mechanical abrasion processes, and so is the salivary film that covers them. Macakova et al. [57] studied wear mechanisms of salivary films formed on model hydrophobic surfaces by means of a mini traction machine. They found that the lubrication properties were irreversibly lost when high loads were applied, a process they attributed to the removal of the adsorbed film. In their experiments, wear was not observed when saliva was present in the bulk, indicating that salivary proteins could rapidly re-adsorb from the bulk reservoir. Moreover, they also observed that wear was more extensive when the pellicle was exposed to liquids of low ionic strengths. They argued that this could be due to a decrease in visco-elasticity and cohesiveness of the film which would result from a loss of salt ion-mediated complexation between the adsorbed protein chains. A different mechanism would be that adhesion between adsorbed components on opposing surfaces may arise due to conformational changes at low ionic strengths. However, this adhesion mechanism could be discarded from the work of Harvey et al. [58] where the lubrication and load-bearing properties of salivary films were studied by means of surface force balance. In this work they showed a net repulsive interaction between salivary films and, in agreement with the work of Nylander et al. [36], a lack of adhesion. This in turn supports the loss of cohesion as the mechanism underlying the increase in wear at low ionic strengths. Harvey et al. [58] also observed that, after performing shear measurements on a spot of a film, the range of normal forces increased. This is indicative of ploughing-up, i.e. the lifting, of material during shear. Ploughing in turn supports the loss of cohesion of the layers at low ionic strengths.

In our group we have developed a novel methodology to measure wear at the nanoscale, and applied it in the study of salivary pellicles [21]. Indeed, the nanoscale is highly relevant as both the size of the components, i.e. mostly proteins, and the dimensions of the pellicle are in the range

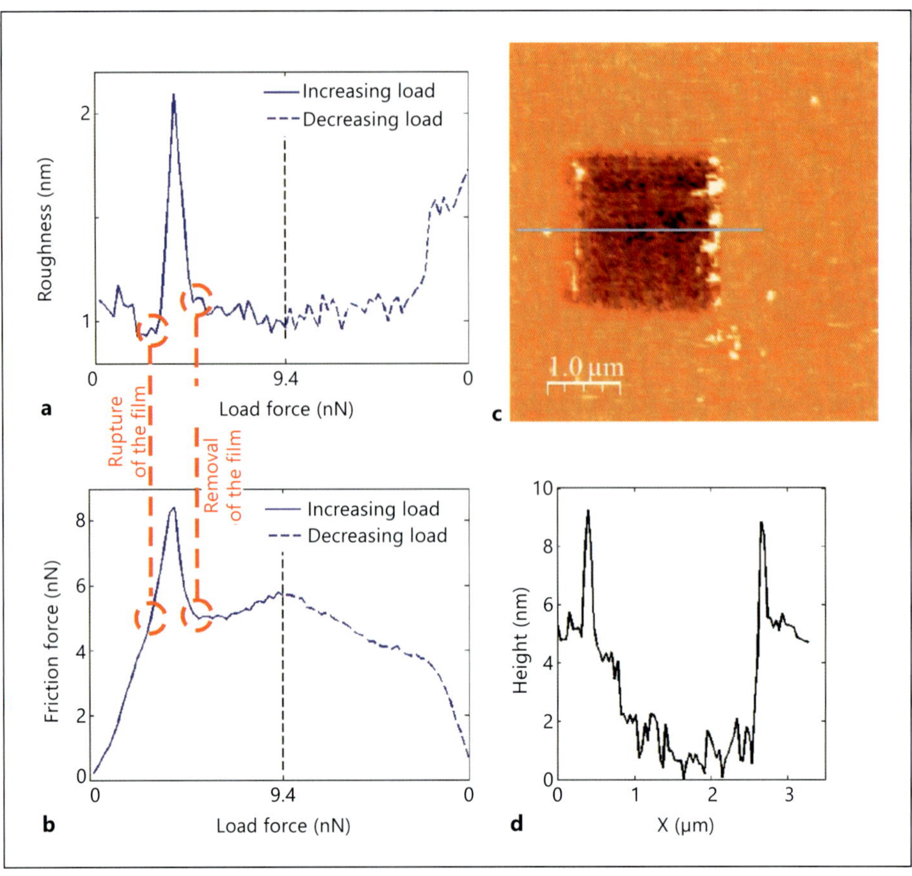

Fig. 3. Nanowear experiment, in water, on a salivary film adsorbed on a model hydrophobic substrate. **a, b** Corresponding roughness and friction plots. **c** Scratched area after the experiment. **d** Cross-sectional profile of the scratched area; X stands for the height profile position along the line drawn in fig. 3c (reprinted with permission from Sotres et al. [21], copyright 2011 American Chemical Society).

of nanometers. The methodology is based on the friction force spectroscopy operation mode of the atomic force microscope (AFM). In this mode, the nanometric tip/probe of the AFM is used to scan micrometric areas of the sample while varying the applied load, i.e. the AFM tip is used to scratch the sample. Simultaneously, the topography of the sample and the friction force that develops between tip and sample are monitored in real time.

Figure 3 shows data from a nanowear measurement, performed in water, of a salivary film formed on a model hydrophobic surface. The load applied while scanning was gradually increased until the film was completely removed, and subsequently gradually lowered. The evolution of the sample topography is illustrated by plotting its roughness in terms of the applied load (roughness plot, fig. 3a). The average friction force between the film and the AFM tip in terms of the applied load is also shown (friction plot, fig. 3b). While increasing the load in the beginning of the experiment, a point is found for which roughness suddenly increases indicating the rupture of the film. Eventually, as the load is further increased, the sample roughness reaches a minimum constant value associated with the complete removal of the pellicle. The forces for which the rupture and removal events occur can be readily determined from the corresponding friction plot and used as indicators of the pellicle strength. This approach has been used to study the influence of the wettability of the substrata on the adsorbed salivary films [22]. Results revealed a drastic increase in the overall strength of salivary films with increasing water wettability of the substrata. These results are especially relevant for understanding mechanical removal processes of dental plaque as the pellicle constitutes its innermost layer. In line with earlier investigations [25], this study suggests that plaque formed on substrata of low wettability would be more prone to separate from the substrata than to exhibit cohesive failure.

Nanowear experiments also provide information on lateral diffusion processes within the pellicle. To illustrate this, we refer again to the experiment shown in figure 3. When after removal of the film the load was decreased, roughness eventually increased indicating the recovery of the surface coverage. This was accompanied by a decrease in friction to values similar to those measured before the rupture. Considering that the experiment was performed in protein-free solution, this suggests that some of the components in the salivary film have the ability to diffuse laterally. Interestingly, the magnitude of neither the sample roughness nor the friction signals did exactly coincide at the beginning and at the end of the scratch. This suggests that the original structure and/or composition of the film were not completely recovered. Moreover, soft imaging of the scratched area (fig. 3c, d) indicated that part of the original film was irreversibly removed. These observations indicate that on hydrophobic surfaces salivary films are composed of two different fractions differing in their ability to diffuse on the substratum.

Nanowear experiments also shed light on the effects of rinsing salivary films with acidic solutions [21]. After rinsing, films formed on hydrophobic substrates became weaker. Moreover, they lacked all the indications (see above) of the presence of non-diffusing components. This suggests that salivary films lost the non-diffusing fraction. For films formed on hydrophilic surfaces, rinsing with acidic solutions resulted in heterogeneous films and formation of big aggregates (typical size of tens of nanometers) separated by zones of clean substrate. The forces needed to remove these aggregates were considerably lower than those needed to remove the films at higher pHs. Therefore, nanowear experiments revealed that rinsing with acidic solutions decreases the strength of salivary films, but apparently by different mechanisms depending on the wettability of the underlying substratum.

References

1 Nieuw Amerongen AV, Bolscher JG, Veerman EC: Salivary proteins: protective and diagnostic value in cariology? Caries Res 2004;38:247–253.

2 Hannig M, Joiner A: The structure, function and properties of the acquired pellicle; in Duckworth RM (ed): The Teeth and Their Environment. Monogr Oral Sci. Basel, Karger, 2006, vol 19, pp 29–64.

3 Nasmyth A: On the structure, physiology, and pathology of the persistent capsular investments and pulp of the tooth. Med Chir Trans 1839;22:310–328.

4 Dawes C, Jenkins GN, Tonge CH: The nomenclature of the integuments of the enamel surface of the teeth. Br Dent J 1963;115:65–68.

5 Hay DI: The adsorption of salivary proteins by hydroxyapatite and enamel. Arch Oral Biol 1967;12:937–946.

6 Mayhall CW: Concerning the composition and source of the acquired enamel pellicle of human teeth. Arch Oral Biol 1970;15:1327–1341.

7 Baier RE, Glantz PO: Characterization of oral in vivo films formed on different types of solid surfaces. Acta Odontol Scand 1978;36:289–301.

8 Arnebrant T: Protein adsorption in the oral environment; in Malmsten M (ed): Biopolymers at Interfaces, ed 2. New York, Dekker, 2003, pp 811–855.

9 Lendenmann U, Grogan J, Oppenheim FG: Saliva and dental pellicle – a review. Adv Dent Res 2000;14:22–28.

10 Hay DI: The interaction of human parotid salivary proteins with hydroxyapatite. Arch Oral Biol 1973;18:1517–1529.

11 Jensen JL, Lamkin MS, Oppenheim FG: Adsorption of human salivary proteins to hydroxyapatite: a comparison between whole saliva and glandular salivary secretions. J Dent Res 1992;71:1569–1576.

12 Vassilakos N, Arnebrant T, Glantz PO: An in vitro study of salivary film formation at solid/liquid interfaces. Scand J Dent Res 1993;101:133–137.

13 Siqueira WL, Custodio W, McDonald EE: New insights into the composition and functions of the acquired enamel pellicle. J Dent Res 2012;91:1110–1118.

14 Hannig M: Ultrastructural investigation of pellicle morphogenesis at two different intraoral sites during a 24-h period. Clin Oral Investig 1999;3:88–95.

15 Iontcheva I, Oppenheim FG, Troxler RF: Human salivary mucin MG1 selectively forms heterotypic complexes with amylase, proline-rich proteins, statherin, and histatins. J Dent Res 1997;76:734–743.

16 Young A, Rykke M, Rolla G: Quantitative and qualitative analyses of human salivary micelle-like globules. Acta Odontol Scand 1999;57:105–110.

17 Vitkov L, Hannig M, Nekrashevych Y, Krautgartner WD: Supramolecular pellicle precursors. Eur J Oral Sci 2004;112:320–325.

18 Arnebrant T, Simonsson T: The effect of ionic surfactants on salivary proteins adsorbed on silica surfaces. Acta Odontol Scand 1991;49:281–288.

19 Lindh L, Arnebrant T, Isberg P, Glantz P: Concentration dependence of adsorption from human whole resting saliva at solid/liquid interfaces: an ellipsometric study. Biofouling 1999;14:189–196.

20 Cardenas M, Elofsson U, Lindh L: Salivary mucin MUC5B could be an important component of in vitro pellicles of human saliva: an in situ ellipsometry and atomic force microscopy study. Biomacromolecules 2007;8:1149–1156.

21 Sotres J, Lindh L, Arnebrant T: Friction force spectroscopy as a tool to study the strength and structure of salivary films. Langmuir 2011;27:13692–13700.

22 Sotres J, Pettersson T, Lindh L, Arnebrant T: Nanowear of salivary films vs substratum wettability. J Dent Res 2012;91:973–978.

23 Barrantes A, Arnebrant T, Lindh L: Characteristics of saliva films adsorbed onto different dental materials studied by QCM-D. Colloids Surf Physicochem Eng Aspects 2014;442:56–62.

24 Scheideler L, Rupp F, Wendel HP, Sathe S, Geis-Gerstorfer J: Photocoupling of fibronectin to titanium surfaces influences keratinocyte adhesion, pellicle formation and thrombogenicity. Dent Mater 2007;23:469–478.

25 Glantz PO: On wettability and adhesiveness. Odontol Rev 1969;20:1–132.

26 Cardenas M, Arnebrant T, Rennie A, Fragneto G, Thomas RK, Lindh L: Human saliva forms a complex film structure on alumina surfaces. Biomacromolecules 2007;8:65–69.

27 Svendsen IE, Arnebrant T, Lindh L: Validation of mechanically assisted sodium dodecyl-sulphate elution as a technique to remove pellicle protein components from human enamel. Biofouling 2008;24:227–233.

28 Svendsen IE, Lindh L: The composition of enamel salivary films is different from the ones formed on dental materials. Biofouling 2009;25:255–261.

29 Santos O, Lindh L, Halthur T, Arnebrant T: Adsorption from saliva to silica and hydroxyapatite surfaces and elution of salivary films by SDS and delmopinol. Biofouling 2010;26:697–710.

30 Norde W: Adsorption of proteins from solution at the solid-liquid interface. Adv Colloid Interface Sci 1986;25:267–340.

31 Malmsten M, Burns N, Veide A: Electrostatic and hydrophobic effects of oligopeptide insertions on protein adsorption. J Colloid Interface Sci 1998;204:104–111.

32 Schupbach P, Oppenheim FG, Lendenmann U, Lamkin MS, Yao Y, Guggenheim B: Electron-microscopic demonstration of proline-rich proteins, statherin, and histatins in acquired enamel pellicles in vitro. Eur J Oral Sci 2001;109:60–68.

33 Hannig M, Khanafer AK, Hoth-Hannig W, Al-Marrawi F, Acil Y: Transmission electron microscopy comparison of methods for collecting in situ formed enamel pellicle. Clin Oral Investig 2005;9:30–37.

34 Hammi AR, Al-Hashimi IH, Nunn ME, Zipp M: Assessment of SS-A and SS-B in parotid saliva of patients with Sjögren's syndrome. J Oral Pathol Med 2005;34:198–203.

35 Hannig C, Ruggeri A, Al-Khayer B, Schmitz P, Spitzmuller B, Deimling D, et al: Electron microscopic detection and activity of glucosyltransferase B, C, and D in the in situ formed pellicle. Arch Oral Biol 2008;53:1003–1010.

36 Nylander T, Arnebrant T, Glantz PO: Interactions between salivary films adsorbed on mica surfaces. Colloids Surf A 1997;129–130:339–344.

37 Lindh L: On the adsorption behaviour of saliva and purified salivary proteins at solid/liquid interfaces. Swed Dent J Suppl 2002;152:1–57.

38 Vroman L, Adams AL: Adsorption of proteins out of plasma and solutions in narrow spaces. J Colloid Interface Sci 1986;111:391–402.

39 Elliott JC: Calcium phosphate biominerals. Rev Mineral Geochem 2002;48:427–453.

40 Nancollas GH: The involvement of calcium phosphates in biological mineralization and demineralization processes. Pure Appl Chem 1992;64:1673–1678.

41 Gibbons R, Houte J: Dental caries. Annu Rev Med 1975;26:121–136.

42 Von Fraunhofer JA, Rogers MM: Dissolution of dental enamel in soft drinks. Gen Dent 2004;52:308–312.

43 Bardow A, Moe D, Nyvad B, Nauntofte B: The buffer capacity and buffer systems of human whole saliva measured without loss of CO_2. Arch Oral Biol 2000;45:1–12.

44 Featherstone J, Behrman J, Bell J: Effect of whole saliva components on enamel demineralization in vitro. Crit Rev Oral Biol Med 1993;4:357–362.

45 Zahradnik R: Modification by salivary pellicles of in vitro enamel remineralization. J Dent Res 1979;58:2066–2073.

46 Moreno E, Varughese K, Hay D: Effect of human salivary proteins on the precipitation kinetics of calcium phosphate. Calcif Tissue Int 1979;28:7–16.

47 Hay D, Carlson E, Schluckebier S, Moreno E, Schlesinger D: Inhibition of calcium phosphate precipitation by human salivary acidic proline-rich proteins: structure-activity relationships. Calcif Tissue Int 1987;40:126–132.

48 Busscher H, White D, Kamminga-Rasker H, Poortinga A, Van der Mei H: Influence of oral detergents and chlorhexidine on soft-layer electrokinetic parameters of the acquired enamel pellicle. Caries Res 2003;37:431–436.

49 Hara AT, Ando M, Gonzalez-Cabezas C, Cury JA, Serra MC, Zero DT: Protective effect of the dental pellicle against erosive challenges in situ. J Dent Res 2006; 85:612–616.

50 Carlen A, Borjesson AC, Nikdel K, Olsson J: Composition of pellicles formed in vivo on tooth surfaces in different parts of the dentition, and in vitro on hydroxyapatite. Caries Res 1998;32: 447–455.

51 Yao Y, Grogan J, Zehnder M, Lendenmann U, Nam B, Wu Z, et al: Compositional analysis of human acquired enamel pellicle by mass spectrometry. Arch Oral Biol 2001;46:293–303.

52 Hu S, Xie Y, Ramachandran P, Ogorzalek Loo RR, Li Y, Loo JA, et al: Large-scale identification of proteins in human salivary proteome by liquid chromatography/mass spectrometry and two-dimensional gel electrophoresis-mass spectrometry. Proteomics 2005;5:1714–1728.

53 Walz A, Stuhler K, Wattenberg A, Hawranke E, Meyer HE, Schmalz G, et al: Proteome analysis of glandular parotid and submandibular-sublingual saliva in comparison to whole human saliva by two-dimensional gel electrophoresis. Proteomics 2006;6:1631–1639.

54 Carlen A, Nikdel K, Wennerberg A, Holmberg K, Olsson J: Surface characteristics and in vitro biofilm formation on glass ionomer and composite resin. Biomaterials 2001;22:481–487.

55 Slomiany BL, Murty VL, Zdebska E, Slomiany A, Gwozdzinski K, Mandel ID: Tooth surface-pellicle lipids and their role in the protection of dental enamel against lactic-acid diffusion in man. Arch Oral Biol 1986;31:187–191.

56 Slomiany BL, Murty VL, Slomiany A: Salivary lipids in health and disease. Prog Lipid Res 1985;24:311–324.

57 Macakova L, Yakubov GE, Plunkett MA, Stokes JR: Influence of ionic strength on the tribological properties of pre-adsorbed salivary films. Tribol Int 2011;44: 956–962.

58 Harvey NM, Yakubov GE, Stokes JR, Klein J: Lubrication and load-bearing properties of human salivary pellicles adsorbed ex vivo on molecularly smooth substrata. Biofouling 2012;28:843–856.

Liselott Lindh
Department of Prosthetic Dentistry
Faculty of Odontology
Malmö University
SE–20506 Malmö (Sweden)
E-Mail liselott.lindh@mah.se

Ligtenberg AJM, Veerman ECI (eds): Saliva: Secretion and Functions.
Monogr Oral Sci. Basel, Karger, 2014, vol 24, pp 40–51 (DOI: 10.1159/000358783)

Antimicrobial Defense Systems in Saliva

Wim van 't Hof · Enno C.I. Veerman · Arie V. Nieuw Amerongen · Antoon J.M. Ligtenberg

Department of Periodontology and Oral Biochemistry, Academic Centre for Dentistry Amsterdam (ACTA), Amsterdam, The Netherlands

Abstract

The oral cavity is one of the most heavily colonized parts of our body. The warm, nutrient-rich and moist environment promotes the growth of a diverse microflora. One of the factors responsible for the ecological equilibrium in the mouth is saliva, which in several ways affects the colonization and growth of bacteria. In this paper, we discuss the various mechanisms by which the composition of the oral microflora is modulated by saliva. Saliva covers the oral hard and soft tissues with a conditioning film which governs the initial attachment of microorganisms, a crucial step in the setup of the oral microflora. It furthermore contains proteins which in the soluble phase bind to bacteria, blocking their adherence to surfaces. When the supply of nutrients is diminished, bacteria use salivary glycoproteins, especially high-molecular-weight mucins, as a source of complex carbohydrates, requiring a consortium of microorganisms for breakdown. In this way saliva promotes the complexity of the oral microflora, which in itself protects against overgrowth by few pathogenic species. Finally, saliva harbors a large panel of antimicrobial proteins which directly and indirectly inhibit uncontrolled outgrowth of bacteria. These include lactoferrin, lactoperoxidase, lysozyme and antimicrobial peptides. Under pathological conditions serum leakage occurs, and saliva mobilizes the humoral and cellular defense mechanisms in the blood. In sum, saliva favors the establishment of a highly diverse microflora, rather than a semisterile environment.

© 2014 S. Karger AG, Basel

The importance of saliva in the maintenance of oral health becomes clear when saliva secretion is disturbed. Patients who suffer from a reduced salivary output are much more susceptible to caries and fungal infections than healthy individuals [1]. When the clearance by saliva is compromised, for instance in intubated patients in intensive care units, within 2 weeks a shift in the oral microflora towards Gram-negative species occurs, often spreading into the respiratory tract and causing pulmonary infections [2]. The total number of bacteria in the oral cavity is estimated to exceed 10^9. Apparently saliva prevents unlimited microbial colonization of the oral tissues, rather than completely eradicating the oral microflora. In this paper we will discuss different mechanisms by which saliva modulates the colonization of oral tissues by microorganisms (table 1):

Table 1. Antimicrobial mechanisms in saliva

Component	Antimicrobial action
Lysozyme	Cell wall degradation
Lactoferrin	Iron depletion, inhibition of biofilm formation
MUC5B	Promotes growth of a complex microflora which suppresses colonization by exogenous microorganisms
MUC7	Microbial agglutination
Lactoperoxidase	Formation of microbicidal products (OSCN⁻)
LL-37, defensins, histatins	Pore formation in microbial membranes
SAG, gp340, DMBT-1	Microbial agglutination
Secretory IgA	Microbial agglutination, immune exclusion
Proline-rich proteins	Microbial adhesion
Statherin	Microbial receptors on dental surface, inhibition of hypha formation of *Candida albicans*

– modulation of bacterial attachment; salivary proteins on hard and soft oral tissues present receptors to which bacteria selectively attach; an overlapping but not identical set of salivary proteins binds to nonadhered microorganisms, promoting their clearance;

– saliva as growth substrate; oral microorganisms are for their growth depending on proteins and glycoproteins in saliva; saliva contains mucins, heavily glycosylated proteins, which are important carbohydrate sources for microorganisms;

– antimicrobial activity; saliva contains a large variety of proteins and peptides which directly and indirectly inhibit growth of microorganisms.

Modulation of Bacterial Attachment

Dental enamel is coated with a film of salivary proteins, the acquired enamel pellicle [3; see the paper by Lindh et al., this vol., pp. 30–39]. Pellicle formation starts with binding of phosphoproteins including proline-rich proteins, statherin and histatins, which have been implicated in mineral homeostasis [4]. Bacteria such as *Streptococcus gordonii*, *Actinomyces naeslundii* and *Porphyromonas gingivalis* and yeasts such as *Candida al-bicans* take advantage of these proteins by using them as a receptor at the dental surface [5–10]. In the maturation phase, a number of other proteins binds, including the large salivary mucin MUC5B. With a molecular weight of $>10^6$ kDa it is by far the largest molecule in saliva. The polypeptide chain of MUC5B is decorated with an extremely heterogeneous set of oligosaccharides, presenting a plethora of structures to which potentially a wide variety of oral microorganisms can bind [11]. Still, in vitro only a very limited set of bacteria has been found which exhibit strong binding to MUC5B, including *Haemophilus parainfluenzae* and *Helicobacter pylori* [12–14]. A possible explanation is that the extreme heterogeneity of the carbohydrate chains leads to a low surface density of each individual binding site on MUC5B. This would make a multivalent binding of bacteria, which is essential for a high avidity adherence, ill-favored or even physically impossible. By preventing bacterial adherence to the dental surfaces, MUC5B could modulate the bacterial colonization of the pellicle (fig. 1). Indeed, attachment to surfaces is the first crucial step for successful colonization in the mouth, since bacteria in the soluble phase will quickly be cleared by mechanical flushing as a result of physiological movements (e.g. swallowing, chewing, speaking) [15]. Besides, by modulation of bacterial adhesion to oral surfaces, in the soluble phase saliva enhances bacterial clearance by promoting agglutination

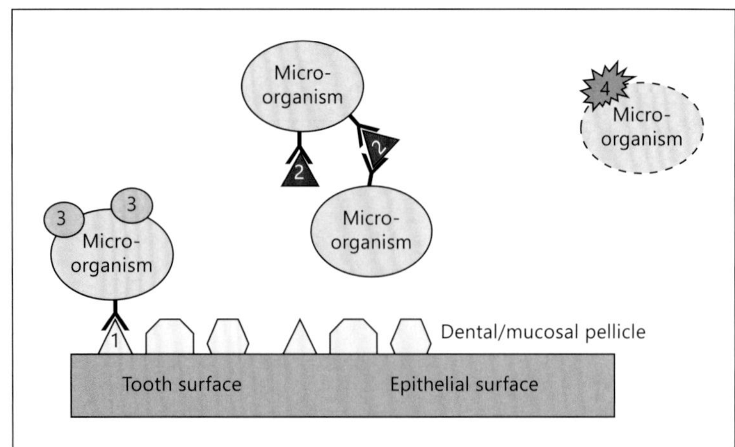

Fig. 1. Effects of saliva on microbial oral colonization. (1) Salivary (glyco)proteins cover the dental and epithelial surfaces forming a pellicle. Microorganisms express specific adhesins enabling them to attach to the salivary receptors on the surface. In this way, microorganisms prevent being flushed away from the oral cavity. (2) Binding of microorganisms to similar receptors in solution prevents adherence to a surface. Multiple receptors on salivary components enable binding of different microorganisms resulting in aggregation. Aggregation is therefore considered a mechanism of bacterial clearance. (3) Salivary components, such as MUC5B, are complex nutrients for microorganisms enabling the development of a complex microbiome. (4) Salivary components, such as histatins and lysozyme, kill microorganisms.

of bacteria and by blocking of bacterial adhesins. The major bacteria-binding proteins in saliva are the low-molecular-weight mucin MUC7, secretory IgA and salivary agglutinin.

The Mucin MUC7
MUC7 [16–18] is a 125-kDa glycoprotein, decorated with short (mainly di- and trisaccharides) oligosaccharide side chains. MUC7 binds and aggregates a wide variety of oral microorganisms including *Streptococcus sanguinis*, *S. mitis*, *S. gordonii*, *Aggregatibacter actinomycetemcomitans* and *Escherichia coli* [19–22]. Binding is mediated by both the carbohydrate side chains of MUC7, in particular sialic acid [20, 21], and unglycosylated peptide domains [22–24]. In addition, MUC7 in a complex with secretory IgA (sIgA) binds to *Staphylococcus aureus* and *Pseudomonas aeruginosa* [25].

Salivary Agglutinin
Salivary agglutinin (SAG), also designated gp340 or DMBT-1, is a glycoprotein of 300–400 kDa [26–29] belonging to the scavenger receptor cysteine-rich superfamily [30, 31]. Initially SAG has been identified as the protein responsible for the *S. mutans*-agglutinating properties of parotid saliva. Later it was shown that SAG, like MUC7, binds a wide variety of bacteria [32]. Its bacteria-binding domain has been mapped to a 16-amino acid loop in its scavenger receptor cysteine-rich domains. Recent studies revealed that SAG exhibits also complement-activating properties [33].

Secretory IgA
sIgA is the most abundant immunoglobulin in saliva. It is secreted by plasma cells in the vicinity of the salivary glands, and subsequently endocytosed by the acinar cells. It is secreted into saliva along

with other salivary proteins. IgA consists of 2 IgA monomers which are covalently linked by a 15-kDa joining chain and with a secretory component coiled around the IgA dimer. The secretory component is heavily glycosylated making sIgA resistant against proteolytic degradation [34]. sIgA binds to a wide variety of bacteria and viruses, preventing their interaction with oral surfaces without triggering an immune response [35]. This mechanism, neutralization of antigens without causing an inflammatory response, is called immune exclusion.

Modulation of Microbial Growth

The maintenance of a benign commensal microflora is considered a defense mechanism in its own right by virtue of its selection pressure on invading exogenous, potentially harmful bacteria. For example *Streptococcus salivarius* produces so-called lantibiotics that inhibit the growth of *Streptococcus pyogenes* [36]. In general, the presence of a highly diverse microbial community prevents outgrowth of a single species, which might lead to a bacterial load exceeding the pathological threshold. The mouth contains various habitats with different physicochemical properties (surface properties, pH, oxidation-reduction potential) which by themselves promote the settlement of a complex microflora. Being an important source of growth substrates for bacteria, saliva also plays an important role in this process. The major carbohydrate sources for microorganisms under conditions when an external supply of nutrients is absent are the large salivary mucins MUC5B. The carbohydrate moiety of MUC5B displays a wide spectrum of oligosaccharide structures, varying in composition, length, branching and acidity. This extremely heterogeneous carbohydrate coat protects MUC5B against enzymatic breakdown, since it requires the combined action of a broad variety of glycosidases, each with a different specificity. In general, a single bacterial species produces a limited set of glycosidases, which can hydrolyze only a small part of the MUC5B carbohydrate moiety. Complete breakdown of mucins requires concerted action of a broad repertoire of enzymes which can only be produced by a consortium of microorganisms [37, 38]. For exogenous microorganisms arriving later on the scene, it is difficult to find their own niche in such a highly interdependent consortium.

Inhibition of Microbial Growth

Saliva contains many compounds that exhibit in vitro antibacterial properties (table 2). It is assumed that their function is to keep the oral microflora within certain limits by preventing excessive colonization of the oral cavity in combination with the bacteria-agglutinating factors. In this paragraph we will focus on the quantitatively most important members of this group.

Lysozyme

Lysozyme, the first discovered antibacterial protein, is found in a variety of mucosal fluids, including tears, saliva and respiratory and cervical secretions. The bactericidal action of lysozyme is generally attributed to its enzymatic breakdown of the bacterial cell wall. Lysozyme hydrolyzes the bond between N-acetylglucosamine and N-acetylmuramic acid of the peptidoglycan moiety. This compromises the cell wall integrity, leading to bacterial cell lysis in hypo-osmotic fluids such as saliva. Interestingly, lysozyme devoid of enzymatic activity, obtained by heat inactivation or site-directed mutagenesis, still exhibits bactericidal activity, suggesting an action independent of its catalytic function [39].

Lactoperoxidase

Peroxidase activity in saliva is derived from two sources: lactoperoxidase from the parotid and submandibular salivary glands, and myeloperoxidase released by polymorphonuclear leukocytes, which migrate into the oral cavity at gingival crevices [40]. The contribution of myeloperoxi-

Table 2. Salivary components for which microorganism-modulating activity has been described

Component	Abbr.	Properties	Ref. No.
α-Amylase		Lipopolysaccharide binding	108, 109
Cystatins		Growth inhibition of *P. gingivalis*	110

Human lipocalin
  Von Ebner's gland protein
  Tear-specific prealbumin
	LCN-1
	VEGh
	TSPA
		Scavenging of oxidation products
		Endonuclease activity
			111, 112
			113

Extraparotid glycoprotein
  Glycoprotein-17
  Gross cystic disease fluid protein
  Prolactin-inducing protein
  Seminal actin-binding protein
	EP-GP
	gp-17
	GDFP-15
	PIP
	SABP
		Bacteria binding in vitro
		Human immunodeficiency virus inhibition in vitro
			114
			115, 116

Chitinase		Cell wall degradation of *C. albicans*	117
Basic proline-rich proteins	bPRPs	Human immunodeficiency virus inhibition in vitro	118
Proline-rich glycoprotein	PRG	Microbial binding	19, 119, 120
Serine leukocyte protease inhibitor	SLPI	Inhibition of *P. aeruginosa* and *C. albicans* in vitro	121
Peptidoglycan recognition proteins	PGRPs PGLYRPs	Bacterial cell wall binding in vitro Activation of bacterial stress response	122, 123
Phospholipase A_2	PLA_2	Hydrolysis of phospholipids	124
Surfactant proteins	SP-A, SP-D	Bacteria binding	125
$α_2$-Macroglobulin	A_2M	Antiviral activity	126
Calprotectin		Anti-*Candida* activity	127
CD14		Lipopolysaccharide binding	128, 129

dase to the total salivary peroxidase activity increases with oral inflammations up to 75%. Salivary peroxidases catalyze the peroxidation of salivary thiocyanate to hypothiocyanate, which has antimicrobial properties, possibly because it inactivates hexokinase, a key enzyme in bacterial glycolysis [41]. Adsorbed onto hydroxyapatite, salivary peroxidase retains its activity, suggesting that it exerts local effects on the dental biofilm. Because of the antimicrobial effects of the lactoperoxidase system, dentifrices and mouthrinses (e.g. Zendium® and Biotène®) have been marketed which enhance the endogenous activity of salivary peroxidase, by supplementing H_2O_2-generating enzyme systems [42].

Lactoferrin

Lactoferrin is present in various mucosal secretions such as tears, milk, and parotid and submandibular saliva. Due to its iron-sequestering properties, it exerts bacteriostatic effects. A correlation was found [43] between the numbers of subgingival *A. actinomycetemcomitans* and the lactoferrin concentrations in saliva of periodontitis patients carrying this microorganism [44, 45]. By sequestering iron ions, lactoferrin induces twitching of bacteria, preventing the buildup of a biofilm [43]. Since microbial biofilms are much more resistant against antimicrobial agents, this property may increase the susceptibility for other antimicrobial substances in saliva.

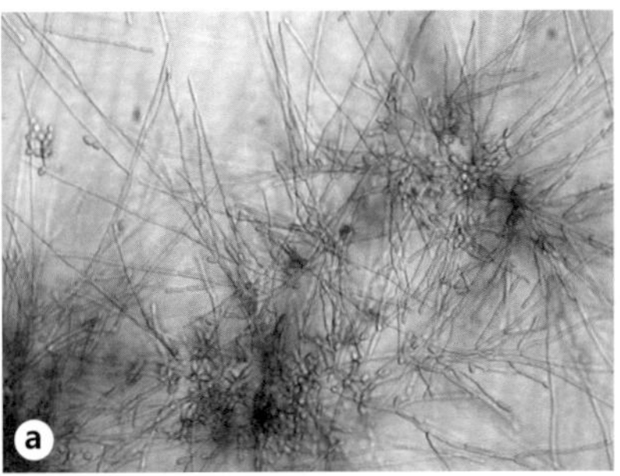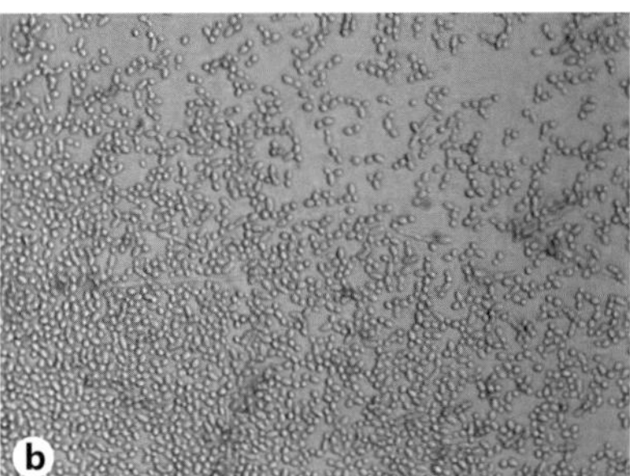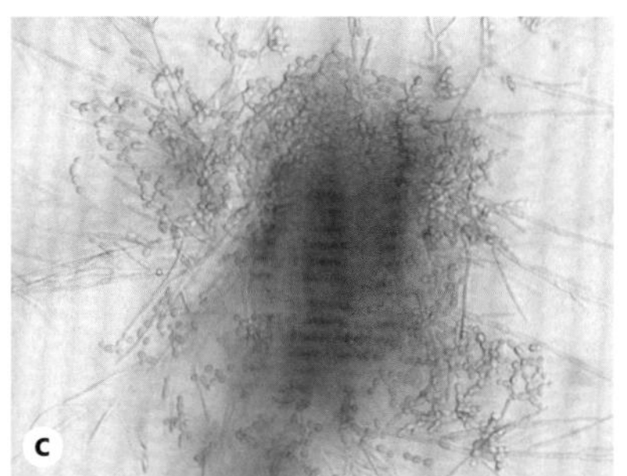

Fig. 2. Effect of saliva and statherin on the hyphal growth of *C. albicans*. *C. albicans* was grown at 37°C in RPMI medium 1:1 diluted with buffer mimicking the ionic composition of saliva (**a**), whole saliva (**b**) and synthetic statherin (**c**) at a final concentration of 20 μM. In RPMI medium at 37°C, *C. albicans* grows as hyphal form which is prevented in the presence of whole saliva and statherin.

Proteolytic degradation of lactoferrin, for example by pepsin in the stomach, may result in the release of several antimicrobial peptides with broad-spectrum activity in vitro and in vivo. One of these peptides, designated lactoferricin, consists of 40 amino acids with 2 distinct bactericidal domains, residues 1–11 and residues 18–31 [45, 46]. For this reason, the antimicrobial activity of lactoferrin in saliva is fundamentally different from that further down the gastrointestinal tract. Apart from its role in noninflammatory defense, lactoferrin is also implicated in inflammatory processes. It is released from the granules of neutrophil granulocytes upon the attack on Gram-negative bacteria in the bloodstream. It functions as chemoattractant for the recruitment of additional neutrophil granulocytes to the site of infection. After binding to lipopolysaccharide from lysed bacteria, it tempers the interleukin-1β- and interleukin-8-mediated inflammation, thus modulating the immune response [47].

Statherin

A completely different antimicrobial mechanism has been discovered for statherin, which inhibits hyphal formation of *C. albicans* [48]. *C. albicans* is a dimorphic fungus which can grow in the globular yeast- of thread-like hyphal form (fig. 2). The yeast form easily disseminates, whereas the hyphal form is able to penetrate tissues and is therefore considered more virulent. Inhibition of hypha formation by statherin might protect against candidiasis by keeping *C. albicans* in the less invasive yeast form. This might explain the found correlation between reduced salivary statherin levels and oral mucositis [49–52].

Antimicrobial Peptides

Antimicrobial peptides with microbicidal activity are widely spread in animals and plants [53]. The general mode of action of these peptides is that, due to their positive charge, they bind the negatively charged surface of microbial membranes. Once they are bound to the surface, they adopt a helical structure resulting in a hydrophobic and positively charged side. With the hydrophobic side peptides integrate in the cell membrane and form a temporary pore through which a rapid lethal efflux of vital cell constituents takes place [54]. Binding and translocation over the cell membrane of these peptides is very sensitive to the ionic strength of the incubation medium [55]. Consequently, their antimicrobial action in hypotonic fluids such as saliva is higher than in serum. The most abundant antimicrobial peptides in saliva are the histatins, the defensins and, to a lesser extent, the cathelicidin LL-37.

Histatins

Salivary histatins are a group of structurally related peptides enriched in the positively charged amino acids histidine, arginine and lysine. These peptides showed potent antimicrobial activity in vitro against *Streptococcus mutans* and *C. albicans* [56, 57]. In human saliva at least 12 histatin peptides have been identified derived from 2 different parent molecules: histatin 1 and histatin 3 [58, 59]. Histatin 5, derived from histatin 3, is the most potent peptide [60, 61]. Although histatins are prone to proteolytic degradation, the active domains appear largely unaffected by the primary cleavage events in whole saliva, suggesting a sustained functional activity of these peptides in the proteolytic environment of the oral cavity [62].

In addition to pore formation, inside the cell, histatin 5 targets the mitochondria, where it binds to Ssa1p. Ssa1p is a member of the human heat shock proteins involved as a chaperone protein in the transport of mitochondrial proteins into the cytosol [63–67]. In *Leishmania* histatin 5 targets mitochondrial ATP synthesis [68].

β-Defensins

β-Defensins (hBDs) are cationic peptides of 30–45 amino acids [69–71]. These are secreted by gingival keratinocytes in the mucosal epithelia and, to a lesser extent, by ductal cells in the salivary glands [72–77]. hBDs display a broad-spectrum antimicrobial activity in vitro to Gram-positive *(S. mutans)* and Gram-negative *(A. actinomycetemcomitans, Fusobacterium nucleatum)* bacteria, fungi *(C. albicans)* and viruses (human immunodeficiency virus, herpes simplex virus) [78–81]. In vitro, synergism between defensins and other antimicrobial innate defense molecules, including lysozyme and lactoferrin, has been demonstrated [82, 83].

In the mouth, hBD-1 is constitutively expressed at low levels, with little regulation in response to infection or other stimuli. hBD-2 and hBD-3 are generally expressed at low levels in vivo under normal conditions, but become induced in the response to microbial colonization and inflammation. Interestingly, some bacterial species, such as *P. gingivalis*, do not induce hBD in vitro [84] while *Treponema denticola* suppresses the induction of hBD-2 and hBD-3 through interaction with their signal transduction pathways [85, 86]. Under inflammatory conditions in vivo, e.g. periodontitis and candidiasis, upregulation of hBD-2 and hBD-3 is correlated with the degree of inflammation (levels of proinflammatory cytokine levels) [87, 88]. Another indication that upregulation of hBD-2 expression is linked to immunological response is that in patients with hyper-IgE syndrome, a form of immunological impairment, low hBD-2 levels correlate with candidiasis incidence [89].

hBDs also display potent chemotactic and immune-signaling properties in vitro, but at 100- to 1,000-fold lower concentrations than their antimicrobial activity [90]. Specifically, defensins can recruit immune cells [91–93], induce degranulation of mast cells and induce release of proinflammatory cytokines [94, 95]. This has raised the question whether, instead of a direct antimicrobial activity of hBDs, regulation of the immune response represents their primary function [96].

Human Cathelicidin LL-37

Cathelicidins are a family of mammalian proteins with a highly conserved N-terminal domain of 102 amino acids and an antimicrobial C-terminal domain varying in length (12–100 amino acids) and composition. In humans, one cathelicidin protein has been identified, designated hCAP18 (human cationic antimicrobial peptide 18 kDa). Cathelicidins are secreted as inactive precursors. The antimicrobial cathelicidin peptide LL-37 is released after proteolytic cleavage by proteinase 3 from neutrophils. LL-37, like other antimicrobial peptides, attacks a wide variety of microorganisms including oral bacteria, *C. albicans* [97, 98] and viruses such as herpes simplex virus. In addition, LL-37 has immunomodulatory properties by, on the one hand, acting as a chemoattractant for neutrophils and, on the other hand, by inhibiting the effect of inflammatory cytokines by lipopolysaccharide binding or by a direct effect on host cells [99].

In the recent literature the capacity for direct microbicidal activity of LL-37 under physiological conditions has been questioned [97, 99–101]. Therefore, the opinion wins ground that the extensive range of immunomodulatory properties in adaptive immunity, that can even counteract the direct microbicidal activity in innate immunity, may represent the major physiological function of LL-37.

Effect of Saliva with Serum and Immune Cells
In everyday practice serum leaks via the sulcus into the oral cavity. The flux of neutrophils through the gingival sulcus together with the abundance of antigen-presenting cells (i.e. Langerhans and dendritic cells) in the oral mucosa suggests that also the cellular immunity may be involved in the maintenance of oral health. This is supported by the observation that inhalation treatment with immunosuppressing corticosteroids e.g. by asthmatic patients often leads to rapid outgrowth of *C. albicans* that was present in the mouth at subclinical levels [102]. This is a clear illustration of the existence of an equilibrium state in the microbial ecology of the mouth, which becomes disrupted when one of the many control systems is impaired.

When SAG is bound to a surface and comes into contact with blood, it activates the lectin pathway of the complement system by binding to mannan-binding lectin [33, 103]. This activity is mediated via a direct interaction with mannan-binding lectin with fucose residues on SAG, although protein epitopes may also be involved [33, 103].

Conclusions

In addition to the salivary proteins described above, a long list of other proteins have been described that somehow affect the oral microflora (table 2). Conclusively we can say that saliva determines in multiple ways the maintenance of a healthy oral cavity and resistance against oral infections. The tolerance of a complex and benign oral flora seems in this context as important as the direct antimicrobial action. The large variety in salivary components and corresponding properties makes it difficult to properly evaluate the physiological relevance of each individual salivary protein [104]. On the one hand, the overlapping properties of several salivary proteins can lead to a high redundancy. For instance, sIgA deficiency in saliva is not uncommon and generally without pathological consequences [105, 106]. On the other hand, in some cases the combination of several salivary proteins may be needed for a functional effect. For example, inhibition of HIV-1 replication by human saliva was completely lost upon fractionation, suggesting that its antiviral activity is the result of a number of salivary components working in concert [107].

References

1 Vissink A, Panders AK, 's-Gravenmade EJ, Vermey A: The causes and consequences of hyposalivation. Ear Nose Throat J 1988;67:166.

2 Dennesen P, van der Ven A, Vlasveld M, Lokker L, Ramsay G, Kessels A, van den Keijbus P, Nieuw Amerongen AV, Veerman ECI: Inadequate salivary flow and poor oral mucosal status in intubated intensive care unit patients. Crit Care Med 2003;31:781–786.

3 Siqueira WL, Custodio W, McDonald EE: New insights into the composition and functions of the acquired enamel pellicle. J Dent Res 2012;91:1110–1118.

4 Schlesinger DH, Hay DI: Complete covalent structure of statherin, a tyrosine-rich acidic peptide which inhibits calcium phosphate precipitation from human parotid saliva. J Biol Chem 1977; 252:1689–1695.

5 Amano A, Kataoka K, Raj PA, Genco RJ, Shizukuishi S: Binding sites of salivary statherin for *Porphyromonas gingivalis* recombinant fimbrillin. Infect Immun 1996;64:4249–4254.

6 Rudney JD, Staikov RK, Johnson JD: Potential biomarkers of human salivary function: a modified proteomic approach. Arch Oral Biol 2009;54:91–100.

7 Cannon RD, Nand AK, Jenkinson HF: Adherence of *Candida albicans* to human salivary components adsorbed to hydroxylapatite. Microbiology 1995;141:213–219.

8 Johansson I, Bratt P, Hay DI, Schluckebier S, Strømberg N: Adhesion of *Candida albicans,* but not *Candida krusei,* to salivary statherin and mimicking host molecules. Oral Microbiol Immunol 2000;15:112–118.

9 Gibbons RJ, Hay DI, Cisar JO, Clark WB: Adsorbed salivary proline-rich protein-1 and statherin – receptors for type-1 fimbriae of *Actinomyces viscosus* T14V-J1 on apatitic surfaces. Infect Immun 1988;56:2990–2993.

10 Kataoka K, Amano A, Kuboniwa M, Horie H, Nagata H, Shizukuishi S: Active sites of salivary proline-rich protein for binding to *Porphyromonas gingivalis* fimbriae. Infect Immun 1997;65:3159–3164.

11 Karlsson NG, Schulz BL, Packer NH: Structural determination of neutral O-linked oligosaccharide alditols by negative ion LC-electronspray-MSn. J Am Soc Mass Spectrom 2004;15:659–672.

12 Veerman ECI, Ligtenberg AJM, Schenkels LPCM, Walgreen-Weterings E, Nieuw Amerongen AV: Binding of human high-molecular weight salivary mucins (MG1) to *Haemophilus parainfluenzae.* J Dent Res 1995;74:351–357.

13 Veerman ECI, Bank CMC, Namavar F, Appelmelk BJ, Bolscher JGM, Nieuw Amerongen AV: Sulfated glycans on oral mucin as receptors for *Helicobacter pylori.* Glycobiology 1997;7:737–743.

14 Bosch JA, de Geus EJ, Ligtenberg TJ, Nazmi K, Veerman EC, Hoogstraten J, Nieuw Amerongen AV: Salivary MUC5B-mediated adherence (ex vivo) of *Helicobacter pylori* during acute stress. Psychosom Med 2000;62:40–49.

15 Dawes C: How much saliva is enough for avoidance of xerostomia? Caries Res 2004;38:236–240.

16 Levine MJ, Reddy MS, Tabak LA, Loomis RE, Bergey EJ, Jones PC, Cohen RE, Stinson MW, Al-Hashimi I: Structural aspects of salivary glycoproteins. J Dent Res 1987;66:436–441.

17 Tabak LA: Structure and function of human salivary mucins. Crit Rev Oral Biol Med 1990;1:229–234.

18 Bolscher JG, Groenink J, van der Kwaak JS, van den Keijbus PA, van 't Hof W, Veerman EC, Nieuw Amerongen AV: Detection and quantification of MUC7 in submandibular, sublingual, palatine, and labial saliva by anti-peptide antiserum. J Dent Res 1999;78:1362–1369.

19 Murray PA, Prakobphol A, Lee T, Hoover CI, Fisher SJ: Adherence of oral streptococci to salivary glycoproteins. Infect Immun 1992;60:31–38.

20 Reddy MS, Levine MJ, Paranchych W: Low-molecular-mass human salivary mucin, MG2: structure and binding of *Pseudomonas aeruginosa.* Crit Rev Oral Biol Med 1993;4:315–323.

21 Groenink J, Ligtenberg AJM, Veerman ECI, Bolscher JGM, Nieuw Amerongen AV: Interaction of the salivary low-molecular-weight mucin (MG2) with *Actinobacillus actinomycetemcomitans.* Antonie van Leeuwenhoek 1996;70:79–87.

22 Moshier A, Reddy MS, Scannapieco FA: Role of type 1 fimbriae in the adhesion of *Escherichia coli* to salivary mucin and secretory immunoglobulin A. Curr Microbiol 1996;33:200–208.

23 Liu B, Rayment S, Oppenheim FG, Troxler RF: Isolation of human salivary mucin MG2 by a novel method and characterization of its interactions with oral bacteria. Arch Biochem Biophys 1999;364:286–293.

24 Liu B, Rayment SA, Gyurko C, Oppenheim FG, Offner GD, Troxler RF: The recombinant N-terminal region of human salivary mucin MG2 (MUC7) contains a binding domain for oral streptococci and exhibits candidacidal activity. Biochem J 2000;345:557–564.

25 Biesbrock AR, Reddy MS, Levine MJ: Interaction of a salivary mucin-secretory immunoglobulin A complex with mucosal pathogens. Infect Immun 1991;59:3492–3497.

26 Holmskov U, Mollenhauer J, Madsen J, Vitved L, Grønlund J, Tornoe I, Kliem A, Reid KBM, Poustka A, Skjødt K: Cloning of GP-340, a putative opsonin receptor for lung surfactant protein D. Proc Natl Acad Sci USA 1999;96:10794–10799.

27 Prakobphol A, Xu F, Hoang VM, Larsson T, Bergström J, Johansson I, Frangsmyr L, Holmskov U, Leffler H, Nilsson C, Boren T, Wright JR, Strømberg N, Fisher SJ: Salivary agglutinin, which binds *Streptococcus mutans* and *Helicobacter pylori,* is the lung scavenger receptor cysteine-rich protein gp-340. J Biol Chem 2000;275:39860–39866.

28 Ligtenberg TJM, Bikker FJ, Groenink J, Tornoe I, Leth-Larsen R, Veerman ECI, Nieuw Amerongen AV, Holmskov U: Human salivary agglutinin binds to lung surfactant protein-D and is identical with scavenger receptor protein gp-340. Biochem J 2001;359:243–248.

29 Ligtenberg AJ, Veerman EC, Nieuw Amerongen AV, Mollenhauer J: Salivary agglutinin/glycoprotein-340/DMBT1: a single molecule with variable composition and with different functions in infection, inflammation and cancer. Biol Chem 2007;388:1275–1289.

30 Martinez VG, Moestrup SK, Holmskov U, Mollenhauer J, Lozano F: The conserved scavenger receptor cysteine-rich superfamily in therapy and diagnosis. Pharmacol Rev 2011;63:967–1000.

31 Sarrias MR, Grønlund J, Padilla O, Madsen J, Holmskov U, Lozano F: The scavenger receptor cysteine-rich (SRCR) domain: an ancient and highly conserved protein module of the innate immune system. Crit Rev Immunol 2004;24:1–37.

32 Ericson T, Rundegren J: Characterization of a salivary agglutinin reacting with a serotype c strain of *Streptococcus mutans.* Eur J Biochem 1983;133:255–261.

33 Leito JT, Ligtenberg AJ, van Houdt M, van den Berg TK, Wouters D: The bacteria binding glycoprotein salivary agglutinin (SAG/gp340) activates complement via the lectin pathway. Mol Immunol 2011;49:185–190.

34 Corthesy B: Role of secretory immunoglobulin A and secretory component in the protection of mucosal surfaces. Future Microbiol 2010;5:817–829.

35 Brandtzaeg P: Secretory immunity with special reference to the oral cavity. J Oral Microbiol 2013, Epub ahead of print.

36 Upton M, Tagg JR, Wescombe P, Jenkinson HF: Intra- and interspecies signaling between *Streptococcus salivarius* and *Streptococcus pyogenes* mediated by SalA and SalA1 lantibiotic peptides. J Bacteriol 2001;183:3931–3938.

37 Kolenbrander PE: Multispecies communities: interspecies interactions influence growth on saliva as sole nutritional source. Int J Oral Sci 2011;3:49–54.

38 Van der Hoeven JS, van den Kieboom CWA, Camp PJM: Utilization of mucin by oral *Streptococcus* species. Antonie van Leeuwenhoek 1990;57:165–172.

39 Ibrahim HR, Matsuzaki T, Aoki T: Genetic evidence that antibacterial activity of lysozyme is independent of its catalytic function. FEBS Lett 2001;506:27–32.

40 Ihalin R, Loimaranta V, Tenovuo J: Origin, structure, and biological activities of peroxidases in human saliva. Arch Biochem Biophys 2006;445:261–268.

41 Ashby MT: Inorganic chemistry of defensive peroxidases in the human oral cavity. J Dent Res 2008;87:900–914.

42 Alves MB, Motta AC, Messina WC, Migliari DA: Saliva substitute in xerostomic patients with primary Sjögren's syndrome: a single-blind trial. Quintessence Int 2004;35:392–396.

43 Singh PK, Parsek MR, Greenberg EP, Welsh MJ: A component of innate immunity prevents bacterial biofilm development. Nature 2002;417:552–555.

44 Groenink J, Walgreen-Weterings J, Nazmi K, Bolscher JGM, Winkelhoff AJ, Nieuw Amerongen AV: Salivary lactoferrin and low M_r mucin MG2 in *Actinobacillus actinomycetemcomitans*-associated periodontitis. J Clin Periodontol 2000;26:269–275.

45 Groenink J, Walgreen-Weterings E, van 't Hof W, Veerman ECI, Nieuw Amerongen AV: Cationic amphipatic peptides, derived from bovine and human lactoferrins, with antimicrobial activity against oral pathogens. FEMS Microbiol Lett 1999;179:217–222.

46 Bellamy W, Takase M, Yamauchi K, Wakabayashi H, Kawase K, Tomita M: Identification of the bactericidal domain of lactoferrin. Biochim Biophys Acta 1992;1121:130–136.

47 Embleton ND, Berrington JE, McGuire W, Stewart CJ, Cummings SP: Lactoferrin: antimicrobial activity and therapeutic potential. Semin Fetal Neonatal Med 2013, Epub ahead of print.

48 Leito JT, Ligtenberg AJ, Nazmi K, Veerman EC: Identification of salivary components that induce transition of hyphae to yeast in *Candida albicans*. FEMS Yeast Res 2009;9:1102–1110.

49 Bencharit S, Altarawneh SK, Baxter SS, Carlson J, Ross GF, Border MB, Mack CR, Byrd WC, Dibble CF, Barros S, Loewy Z, Offenbacher S: Elucidating role of salivary proteins in denture stomatitis using a proteomic approach. Mol Biosyst 2012;8:3216–3223.

50 Isola M, Lantini M, Solinas P, Diana M, Isola R, Loy F, Cossu M: Diabetes affects statherin expression in human labial glands. Oral Dis 2011;17:685–689.

51 Isola M, Solinas P, Proto E, Cossu M, Lantini MS: Reduced statherin reactivity of human submandibular gland in diabetes. Oral Dis 2011;17:217–220.

52 Isola M, Cossu M, Diana M, Isola R, Loy F, Solinas P, Lantini MS: Diabetes reduces statherin in human parotid: immunogold study and comparison with submandibular gland. Oral Dis 2012;18:360–364.

53 Zasloff M: Antimicrobial peptides of multicellular organisms. Nature 2002;415:389–395.

54 Mochon AB, Liu H: The antimicrobial peptide histatin-5 causes a spatially restricted disruption on the *Candida albicans* surface, allowing rapid entry of the peptide into the cytoplasm. PLoS Pathog 2008;4:e1000190.

55 Helmerhorst EJ, van 't Hof W, Veerman EC, Simoons-Smit I, Nieuw Amerongen AV: Synthetic histatin analogues with broad-spectrum antimicrobial activity. Biochem J 1997;326:39–45.

56 Pollock JJ, Denepitiya L, MacKay BJ, Iacono VJ: Fungistatic and fungicidal activity of human parotid salivary histidine-rich polypeptides on *Candida albicans*. Infect Immun 1984;44:702–707.

57 MacKay BJ, Denepitiya L, Iacono VJ, Krost SB, Pollock JJ: Growth-inhibitory and bactericidal effects of human parotid salivary histidine-rich polypeptides on *Streptococcus mutans*. Infect Immun 1984;44:695–701.

58 Sabatini LM, He YZ, Azen EA: Structure and sequence determination of the gene encoding human salivary statherin. Gene 1990;89:245–251.

59 Van der Spek JC, Offner GD, Troxler RF, Oppenheim FG: Molecular cloning of human submandibular histatins. Arch Oral Biol 1990;35:137–143.

60 Oppenheim FG, Xu T, McMillian FM, Levitz SM, Diamond RD, Offner GD, Troxler RF: Histatins, a novel family of histidine-rich proteins in human parotid secretion. Isolation, characterization, primary structure, and fungistatic effects on *Candida albicans*. J Biol Chem 1988;263:7472–7477.

61 Oppenheim FG, Yang YC, Diamond RD, Hyslop D, Offner GD, Troxler RF: The primary structure and functional characterization of the neutral histidine-rich polypeptide from human parotid secretion. J Biol Chem 1986;261:1177–1182.

62 Sun X, Salih E, Oppenheim FG, Helmerhorst EJ: Kinetics of histatin proteolysis in whole saliva and the effect on bioactive domains with metal-binding, antifungal, and wound-healing properties. FASEB J 2009;23:2691–2701.

63 Helmerhorst EJ, Breeuwer P, van 't Hof W, Walgreen-Weterings E, Oomen LCJM, Veerman ECI, Nieuw Amerongen AV, Abee T: The cellular target of histatin 5 on *Candida albicans* is the energized mitochondrion. J Biol Chem 1999;274:7286–7291.

64 Edgerton M, Koshlukova SE, Lo TE, Chrzan BG, Straubinger RM, Raj PA: Candidacidal activity of salivary histatins. Identification of a histatin 5-binding protein on *Candida albicans*. J Biol Chem 1998;273:20438–20447.

65 Ruissen ALA, Groenink J, Helmerhorst EJ, Walgreen-Weterings E, van 't Hof W, Veerman EC, Nieuw Amerongen AV: Effects of histatin 5 and derived peptides on *Candida albicans*. Biochem J 2001;356:361–368.

66 Li XS, Sun JN, Okamoto-Shibayama K, Edgerton M: *Candida albicans* cell wall ssa proteins bind and facilitate import of salivary histatin 5 required for toxicity. J Biol Chem 2006;281:22453–22463.

67 Li XS, Reddy MS, Baev D, Edgerton M: *Candida albicans* Ssa1/2p is the cell envelope binding protein for human salivary histatin 5. J Biol Chem 2003;278:28553–28561.

68 Luque-Ortega JR, van 't Hof W, Veerman EC, Saugar JM, Rivas L: Human antimicrobial peptide histatin 5 is a cell-penetrating peptide targeting mitochondrial ATP synthesis in *Leishmania*. FASEB J 2008;22:1817–1828.

69 Harder J, Bartels J, Christophers E, Schröder JM: Isolation and characterization of human beta-defensin-3, a novel human inducible peptide antibiotic. J Biol Chem 2001;276:5707–5713.

70 Harder J, Siebert R, Zhang Y, Matthiesen P, Christophers E, Schlegelberger B, Schröder JM: Mapping of the gene encoding human beta-defensin-2 (DEFB2) to chromosome region 8p22–p23.1. Genomics 1997;46:472–475.

71 Schröder JM, Harder J: Human beta-defensin-2. Int J Biochem Cell Biol 1999;31:645–651.

72 Abiko Y, Saitoh M: Salivary defensins and their importance in oral health and disease. Curr Pharm Des 2007;13:3065–3072.

73 Dunsche A, Acil Y, Siebert R, Harder J, Schröder JM, Jepsen S: Expression profile of human defensins and antimicrobial proteins in oral tissues. J Oral Pathol Med 2001;30:154–158.

74 Gursoy UK, Kononen E, Luukkonen N, Uitto VJ: Human neutrophil defensins and their effect on epithelial cells. J Periodontol 2013;84:126–133.

75 Gursoy UK, Kononen E: Understanding the roles of gingival beta-defensins. J Oral Microbiol 2012, Epub ahead of print.

76 Mathews M, Jia HP, Guthmiller J, Losh G, Graham S, Johnson GK, Tack BF, McCray PB: Production of beta defensin antimicrobial peptides by the oral mucosa and salivary glands. Infect Immun 1999;67:2740–2745.

77 Weinberg A, Krisanaprakornkit S, Dale BA: Epithelial antimicrobial peptides: review and significance for oral applications. Crit Rev Oral Biol Med 1998;9:399–414.

78 Maisetta G, Batoni G, Esin S, Raco G, Bottai D, Favilli F, Florio W, Campa M: Susceptibility of *Streptococcus mutans* and *Actinobacillus actinomycetemcomitans* to bactericidal activity of human beta-defensin 3 in biological fluids. Antimicrob Agents Chemother 2005;49:1245–1248.

79 Mineshiba F, Takashiba S, Mineshiba J, Matsuura K, Kokeguchi S, Murayama Y: Antibacterial activity of synthetic human B defensin-2 against periodontal bacteria. J Int Acad Periodontol 2003;5:35–40.

80 Ouhara K, Komatsuzawa H, Yamada S, Shiba H, Fujiwara T, Ohara M, Sayama K, Hashimoto K, Kurihara H, Sugai M: Susceptibilities of periodontopathogenic and cariogenic bacteria to antibacterial peptides, {beta}-defensins and LL37, produced by human epithelial cells. J Antimicrob Chemother 2005;55:888–896.

81 Weinberg A, Quinones-Mateu ME, Lederman MM: Role of human beta-defensins in HIV infection. Adv Dent Res 2006;19:42–48.

82 Bals R, Wang X, Wu Z, Freeman T, Bafna V, Zasloff M, Wilson JM: Human beta-defensin 2 is a salt-sensitive peptide antibiotic expressed in human lung. J Clin Invest 1998;102:874–880.

83 Maisetta G, Batoni G, Esin S, Luperini F, Pardini M, Bottai D, Florio W, Giuca MR, Gabriele M, Campa M: Activity of human beta-defensin 3 alone or combined with other antimicrobial agents against oral bacteria. Antimicrob Agents Chemother 2003;47:3349–3351.

84 Krisanaprakornkit S, Kimball JR, Weinberg A, Darveau RP, Bainbridge BW, Dale BA: Inducible expression of human beta-defensin 2 by *Fusobacterium nucleatum* in oral epithelial cells: multiple signaling pathways and role of commensal bacteria in innate immunity and the epithelial barrier. Infect Immun 2000;68:2907–2915.

85 Shin J, Choi Y: The fate of *Treponema denticola* within human gingival epithelial cells. Mol Oral Microbiol 2012;27:471–482.

86 Shin JE, Choi Y: *Treponema denticola* suppresses expression of human beta-defensin-2 in gingival epithelial cells through inhibition of TNFalpha production and TLR2 activation. Mol Cells 2010;29:407–412.

87 Pereira AL, Franco GC, Cortelli SC, Aquino DR, Costa FO, Raslan SA, Cortelli JR: Influence of periodontal status and periodontopathogens on levels of oral human beta-defensin-2 in saliva. J Periodontol 2013;84:1445–1453.

88 Sawaki K, Mizukawa N, Yamaai T, Fukunaga J, Sugahara T: Immunohistochemical study on expression of alpha-defensin and beta-defensin-2 in human buccal epithelia with candidiasis. Oral Dis 2002;8:37–41.

89 Conti HR, Baker O, Freeman AF, Jang WS, Holland SM, Li RA, Edgerton M, Gaffen SL: New mechanism of oral immunity to mucosal candidiasis in hyper-IgE syndrome. Mucosal Immunol 2011;4:448–455.

90 Yang D, Biragyn A, Kwak LW, Oppenheim JJ: Mammalian defensins in immunity: more than just microbicidal. Trends Immunol 2002;23:291–296.

91 Yang D, Chertov O, Bykovskaia SN, Chen Q, Buffo MJ, Shogan J, Anderson M, Schröder JM, Wang JM, Howard OM, Oppenheim JJ: Beta-defensins: linking innate and adaptive immunity through dendritic and T cell CCR6. Science 1999;286:525–528.

92 Niyonsaba F, Ogawa H, Nagaoka I: Human beta-defensin-2 functions as a chemotactic agent for tumour necrosis factor-alpha-treated human neutrophils. Immunology 2004;111:273–281.

93 Niyonsaba F, Iwabuchi K, Matsuda H, Ogawa H, Nagaoka I: Epithelial cell-derived human beta-defensin-2 acts as a chemotaxin for mast cells through a pertussis toxin-sensitive and phospholipase C-dependent pathway. Int Immunol 2002;14:421–426.

94 Niyonsaba F, Ushio H, Hara M, Yokoi H, Tominaga M, Takamori K, Kajiwara N, Saito H, Nagaoka I, Ogawa H, Okumura K: Antimicrobial peptides human beta-defensins and cathelicidin LL-37 induce the secretion of a pruritogenic cytokine IL-31 by human mast cells. J Immunol 2010;184:3526–3534.

95 Niyonsaba F, Ushio H, Nakano N, Ng W, Sayama K, Hashimoto K, Nagaoka I, Okumura K, Ogawa H: Antimicrobial peptides human beta-defensins stimulate epidermal keratinocyte migration, proliferation and production of proinflammatory cytokines and chemokines. J Invest Dermatol 2007;127:594–604.

96 Diamond G, Ryan L: Beta-defensins: what are they really doing in the oral cavity? Oral Dis 2011;17:628–635.

97 Den Hertog AL, van Marle J, van Veen HA, van 't Hof W, Bolscher JG, Veerman EC, Nieuw Amerongen AV: Candidacidal effects of two antimicrobial peptides: histatin 5 causes small membrane defects, but LL-37 causes massive disruption of the cell membrane. Biochem J 2005;388:689–695.

98 Den Hertog AL, van Marle J, Veerman EC, Valentijn-Benz M, Nazmi K, Kalay H, Grun CH, van 't Hof W, Bolscher JG, Nieuw Amerongen AV: The human cathelicidin peptide LL-37 and truncated variants induce segregation of lipids and proteins in the plasma membrane of *Candida albicans*. Biol Chem 2006;387:1495–1502.

99 Barlow PG, Beaumont PE, Cosseau C, Mackellar A, Wilkinson TS, Hancock RE, Haslett C, Govan JR, Simpson AJ, Davidson DJ: The human cathelicidin LL-37 preferentially promotes apoptosis of infected airway epithelium. Am J Respir Cell Mol Biol 2010;43:692–702.

100 Bowdish DM, Davidson DJ, Scott MG, Hancock RE: Immunomodulatory activities of small host defense peptides. Antimicrob Agents Chemother 2005;49:1727–1732.

101 Pompilio A, Scocchi M, Pomponio S, Guida F, Di Primio A, Fiscarelli E, Gennaro R, Di Bonaventura G: Antibacterial and anti-biofilm effects of cathelicidin peptides against pathogens isolated from cystic fibrosis patients. Peptides 2011;32:1807–1814.

102 Van Boven JF, de Jong-van den Berg LT, Vegter S: Inhaled corticosteroids and the occurrence of oral candidiasis: a prescription sequence symmetry analysis. Drug Saf 2013;36:231–236.

103 Reichhardt MP, Loimaranta V, Thiel S, Finne J, Meri S, Jarva H: The salivary scavenger and agglutinin binds MBL and regulates the lectin pathway of complement in solution and on surfaces. Front Immunol 2012;3:205.

104 Nieuw Amerongen AV, Veerman EC: Saliva – the defender of the oral cavity. Oral Dis 2002;8:12–22.

105 Fernandes FR, Nagao AT, Mayer MP, Zelante F, Carneiro-Sampaio MM: Compensatory levels of salivary IgM anti-*Streptococcus mutans* antibodies in IgA-deficient patients. J Investig Allergol Clin Immunol 1995;5:151–155.

106 Slack E, Hapfelmeier S, Stecher B, Velykoredko Y, Stoel M, Lawson MA, Geuking MB, Beutler B, Tedder TF, Hardt WD, Bercik P, Verdu EF, McCoy KD, Macpherson AJ: Innate and adaptive immunity cooperate flexibly to maintain host-microbiota mutualism. Science 2009;325:617–620.

107 Bolscher JG, Nazmi K, Ran LJ, van Engelenburg FA, Schuitemaker H, Veerman EC, Nieuw Amerongen AV: Inhibition of HIV-1 IIIB and clinical isolates by human parotid, submandibular, sublingual and palatine saliva. Eur J Oral Sci 2002;110:149–156.

108 Baik JE, Hong SW, Choi S, Jeon JH, Park OJ, Cho K, Seo DG, Kum KY, Yun CH, Han SH: Alpha-amylase is a human salivary protein with affinity to lipopolysaccharide of *Aggregatibacter actinomycetemcomitans*. Mol Oral Microbiol 2013;28:142–153.

109 Choi S, Baik JE, Jeon JH, Cho K, Seo DG, Kum KY, Yun CH, Han SH: Identification of *Porphyromonas gingivalis* lipopolysaccharide-binding proteins in human saliva. Mol Immunol 2011;48:2207–2213.

110 Blankenvoorde MFJ, van 't Hof W, Walgreen-Weterings E, van Steenbergen TJMBH, Veerman ECI, Nieuw Amerongen AV: Cystatin and cystatin-derived peptides have antibacterial activity against the pathogen *Porphyromonas gingivalis*. Biol Chem 1998;379:1371–1375.

111 Redl B: Human tear lipocalin. Biochim Biophys Acta 2000;1482:241–248.

112 Lechner M, Wojnar P, Redl B: Human tear lipocalin acts as an oxidative-stress-induced scavenger of potentially harmful lipid peroxidation products in a cell culture system. Biochem J 2001;356:129–135.

113 Yusifov TN, Abduragimov AR, Gasymov OK, Glasgow BJ: Endonuclease activity in lipocalins. Biochem J 2000;347:815–819.

114 Schenkels LC, Walgreen-Weterings E, Oomen LC, Bolscher JG, Veerman EC, Nieuw Amerongen AV: In vivo binding of the salivary glycoprotein EP-GP (identical to GCDFP-15) to oral and non-oral bacteria detection and identification of EP-GP binding species. Biol Chem 1997;378:83–88.

115 Autiero M, Gaubin M, Mani JC, Castejon C, Martin MES, Guardiola J, Piatiertonneau D: Surface plasmon resonance analysis of gp17, a natural CD4 ligand from human seminal plasma inhibiting human immunodeficiency virus type-1 gp120-mediated syncytium formation. Eur J Biochem 1997;245:208–213.

116 Gaubin M, Auteiro M, Basmaciogullari S, Metivier D, Misehal Z, Culerrier R, Oudin A, Guardiola J, Piatier-Tonneau D: Potent inhibition of CD4/TCR-mediated T cell apoptosis by a CD-4 binding glycoprotein secreted from breast tumor and seminal vessicle. J Immunol 1999;162:2631–2638.

117 Van Steijn GJ, Nieuw Amerongen AV, Veerman EC, Kasanmoentalib S, Overdijk B: Chitinase in whole and glandular human salivas and in whole saliva of patients with periodontal inflammation. Eur J Oral Sci 1999;107:328–337.

118 Robinovitch MR, Ashley RL, Iversen JM, Vigoren EM, Oppenheim FG, Lamkin M: Parotid salivary basic proline-rich proteins inhibit HIV-I infectivity. Oral Dis 2001;7:86–93.

119 Walz A, Odenbreit S, Stuhler K, Wattenberg A, Meyer HE, Mahdavi J, Boren T, Ruhl S: Identification of glycoprotein receptors within the human salivary proteome for the lectin-like BabA and SabA adhesins of *Helicobacter pylori* by fluorescence-based 2-D bacterial overlay. Proteomics 2009;9:1582–1592.

120 Walz A, Odenbreit S, Mahdavi J, Boren T, Ruhl S: Identification and characterization of binding properties of *Helicobacter pylori* by glycoconjugate arrays. Glycobiology 2005;15:700–708.

121 Chattopadhyay A, Gray LR, Patton LL, Caplan DJ, Slade GD, Tien HC, Shugars DC: Salivary secretory leukocyte protease inhibitor and oral candidiasis in human immunodeficiency virus type 1-infected persons. Infect Immun 2004;72:1956–1963.

122 Dziarski R, Kashyap DR, Gupta D: Mammalian peptidoglycan recognition proteins kill bacteria by activating two-component systems and modulate microbiome and inflammation. Microb Drug Resist 2012;18:280–285.

123 Dziarski R, Gupta D: Review: mammalian peptidoglycan recognition proteins (PGRPs) in innate immunity. Innate Immun 2010;16:168–174.

124 Fonseca-Maldonado R, Ferreira TL, Ward RJ: The bacterial effect of human secreted group IID phospholipase A_2 results from both hydrolytic and non-hydrolytic activities. Biochimie 2012;94:1437–1440.

125 Brauer L, Moschter S, Beileke S, Jager K, Garreis F, Paulsen FP: Human parotid and submandibular glands express and secrete surfactant proteins A, B, C and D. Histochem Cell Biol 2009;132:331–338.

126 Chen CH, Zhang XQ, Lo CW, Liu PF, Liu YT, Gallo RL, Hsieh MF, Schooley RT, Huang CM: The essentiality of alpha-2-macroglobulin in human salivary innate immunity against new H1N1 swine origin influenza A virus. Proteomics 2010;10:2396–2401.

127 Sweet SP, Denbury AN, Challacombe SJ: Salivary calprotectin levels are raised in patients with oral candidiasis or Sjögren's syndrome but decreased by HIV infection. Oral Microbiol Immunol 2001;16:119–123.

128 Sugawara S, Uehara A, Tamai R, Takada H: Innate immune responses in oral mucosa. J Endotoxin Res 2002;8:465–468.

129 Uehara A, Sugawara S, Watanabe K, Echigo S, Sato M, Yamaguchi T, Takada H: Constitutive expression of a bacterial pattern recognition receptor, CD14, in human salivary glands and secretion as a soluble form in saliva. Clin Diagn Lab Immunol 2003;10:286–292.

Arie V. Nieuw Amerongen
Department of Periodontology and Oral Biochemistry
Academic Centre of Dentistry Amsterdam ACTA
Gustav Mahlerlaan 3004, NL–1081 LA Amsterdam (The Netherlands)
E-Mail A.vannieuwamerongen@kliksafe.nl

Ligtenberg AJM, Veerman ECI (eds): Saliva: Secretion and Functions.
Monogr Oral Sci. Basel, Karger, 2014, vol 24, pp 52–60 (DOI: 10.1159/000358784)

Saliva and Wound Healing

Henk S. Brand[a, b] · Antoon J.M. Ligtenberg[a] · Enno C.I. Veerman[a]

Departments of [a]Periodontology and Oral Biochemistry and [b]Oral-Maxillofacial Surgery, Academic Centre
for Dentistry Amsterdam (ACTA), Amsterdam, The Netherlands

Abstract

Oral wounds heal faster and with less scar formation than
skin wounds. One of the key factors involved is saliva,
which promotes wound healing in several ways. Saliva
creates a humid environment, thus improving the survival and functioning of inflammatory cells that are crucial for wound healing. In addition, saliva contains several proteins which play a role in the different stages of
wound healing. Saliva contains substantial amounts of
tissue factor, which dramatically accelerates blood clotting. Subsequently, epidermal growth factor in saliva promotes the proliferation of epithelial cells. Secretory leucocyte protease inhibitor inhibits the tissue-degrading
activity of enzymes like elastase and trypsin. Absence of
this protease inhibitor delays oral wound healing. Salivary histatins in vitro promote wound closure by enhanc-
ing cell spreading and cell migration, but do not stimulate cell proliferation. A synthetic cyclic variant of histatin
exhibits a 1,000-fold higher activity than linear histatin,
which makes this cyclic variant a promising agent for the
development of a new wound healing medication. Conclusively, recognition of the many roles salivary proteins
play in wound healing makes saliva a promising source
for the development of new drugs involved in tissue regeneration. © 2014 S. Karger AG, Basel

Every day the oral mucosa is challenged by mechanical forces and chemical stress during activities such as eating, drinking, biting, chewing and
speaking. The oral epithelium provides limited
protection against these forces. However, the oral
tissues are covered with hydrophilic mucous layers. The mucous layers lubricate the teeth and mucosa, offering protection against frictional forces.
In addition, the mucous layers protect the oral epithelium against desiccation, toxic substances and
microbial invasion. Salivary mucins are the main
constituents of the hydrophilic mucous layers.

Despite the protection provided by the mucous layers, damage to the oral tissues sometimes
occurs and needs to be healed. As dentists may
observe in their practice, wounds in the oral cavity heal faster and with less scar formation than
skin wounds. An important difference between
the skin and oral mucosa is the presence of saliva,
which most likely plays a role in wound healing.
Saliva is already 'therapeutically' used during the
instinctive licking of wounds, which is not only
observed in humans, but also in monkeys, cats,
dogs, horses and other animals. There have even
been reports that Fijian fishermen allow dogs to
lick their wounds to promote wound healing [1].

Healing of Wounds in the Skin and the Oral Mucosa

The healing of skin wounds is characterized by several, partly overlapping, stages:

(1) hemostasis – immediately after injury, several processes will begin to limit further blood loss; these include vasoconstriction, platelet plug formation and blood coagulation;

(2) the inflammatory phase – macrophages and other inflammatory cells will remove bacteria and necrotic cell debris; these inflammatory cells also secrete factors that stimulate division and migration of cells such as epithelial cells and fibroblasts;

(3) the proliferative phase – in this phase regeneration takes place; this includes angiogenesis, deposition of a new collagen matrix and formation of granulation tissue; wound contraction may occur, and the wound is gradually covered with epithelial cells;

(4) remodelling phase – in this final phase, remodelling of the collagen network takes place, and cells that are no longer needed are removed by apoptosis.

Successful completion of wound healing may take between 1 month and more than 2 years. Wound healing in the oral cavity follows the same phases as the healing of the skin, but there is a clear difference. Intra-oral wounds heal faster and with less scar formation than wounds of the skin, which is illustrated by the fact that extraction sockets usually heal rapidly without complications. The rapid healing of intra-oral wounds was also confirmed in a model study with pigs, whose skin closely resembles the human skin. Comparable surgical wounds were made in the palatal mucosa and in the skin. After 14 days, the palatal wounds were clinically closed, and after 28 days the original location of the wound could hardly be recognized from the surrounding unaffected tissue (fig. 1). In contrast, skin wounds were still covered with a crust after 14 days, and after 28 days the original wound could still easily

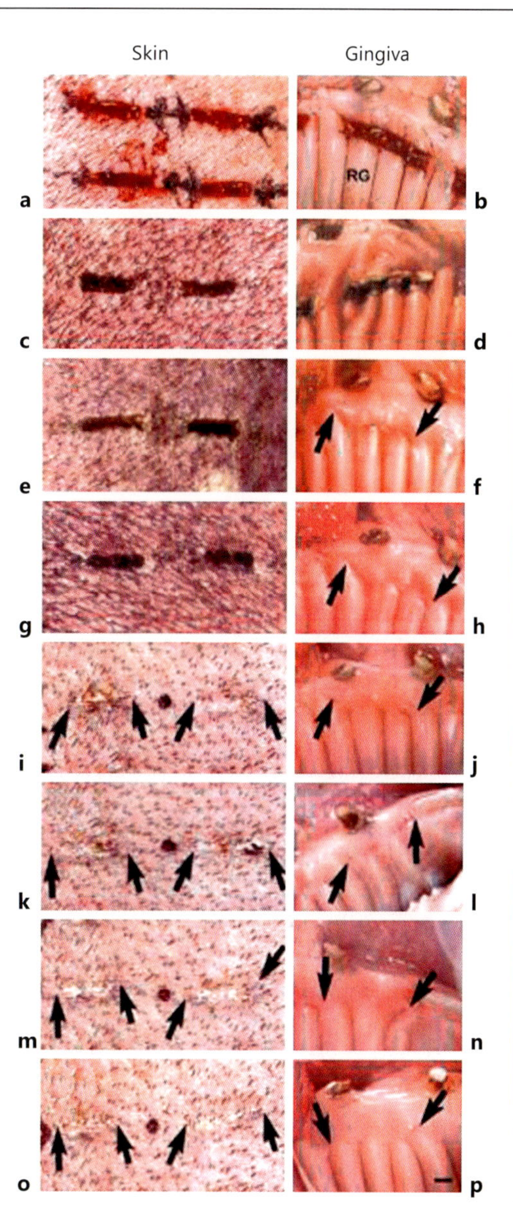

Fig. 1. Clinical scar formation of standardized wounds in pig skin (left column) and pig palatal mucosa (right column), from the time of wounding (**a, b**) and 3 (**c, d**), 7 (**e, f**), 14 (**g, h**), 28 (**i, j**), 35 (**k, l**), 42 (**m, n**) and 49 days (**o, p**) after wounding. Arrows indicate the location of the original wound margins (from Wong et al. [2]).

be recognized [2]. In a study with human volunteers, the size of standard circular wounds in the hard palatal mucosa decreased very rapidly, especially in male and younger individuals [3].

Several factors play a role in the more rapid wound healing of the oral mucosa. First, the oral mucosa has a higher turnover of cells than the skin, enabling a faster repair. Second, the oral mucosa is highly vascularized, thus facilitating the rapid recruitment of inflammatory cells, growth factors and nutrients to the wound. In a later stage, a high degree of vascularization will also facilitate the removal of phagocytosed bacteria and necrotic cells. Third, the humid environment in the mouth also promotes wound healing. The humid atmosphere in the oral cavity prevents dehydration and associated cell death, and improves the supply of nutrients. Wound healing is a complicated process in which many cells are involved, such as neutrophils, macrophages, epithelial cells and fibroblasts. The high survival and improved nutrient supply of these cells in a humid environment will accelerate re-epithelialization, as epithelial cells migrate faster on the humid surface of a wound than under a dry crust.

In addition to the factors mentioned above, saliva itself will promote oral wound healing. Saliva not only creates the humid conditions in the oral cavity, but it also contains several proteins and peptides that enhance the wound healing process.

Haemostasis

Wound healing begins with haemostasis, the arrest of bleeding, necessary before tissue repair can start. Immediately after damage of a blood vessel, platelets come into contact with the underlying connective tissue, aggregate and form a plug to prevent further blood loss. As soon as blood leaks from a vessel, it will come into contact with tissue factor, a protein expressed by smooth muscle cells and subendothelial cells. This tissue factor will activate the coagulation cascade. The coagulation cascade encompasses a series of inactive pro-enzymes, the coagulation factors. The final step of the coagulation cascade is the conversion of fibrinogen into fibrin resulting in the deposition of an insoluble fibrin network on the aggregated platelets.

Already in the 1920s and 1930s, several studies showed that addition of small amounts of saliva to blood accelerated blood clotting (fig. 2) [4–6]. Recently it was shown that saliva is a rich source of tissue factor. Tissue factor in saliva is bound to the membrane of exosomes, small membrane-encapsulated particles with a diameter of 30–90 nm. The exosomes in saliva are derived from epithelial cells and are released when the membranes of multivesicular bodies fuse with the plasma membrane [7].

Antimicrobial Activity

Not all the conditions in the oral cavity are beneficial for wound healing. The warm, humid and nutrient-rich environment is optimal for the growth and development of bacteria. The oral cavity harbours a very complex microbiome comprising more than 1,000 different species of bacteria, fungi and viruses [8]. The total number of bacteria in saliva is estimated to be 10^8–10^9/ml. This microbiome is well able to infect wounds as is clearly demonstrated in animal bite wounds [9]. Mucosal wounds may also become infected. As the presence of bacteria initiates an inflammatory response, this will delay wound healing. In response to bacteria and bacterial toxins, inflammatory cells will migrate to the damaged area. After arrival at the damaged area, the inflammatory cells will secrete interleukins and proteolytic enzymes which may considerably delay wound healing [10].

Saliva contains a wide variety of proteins and peptides that protect against microbial

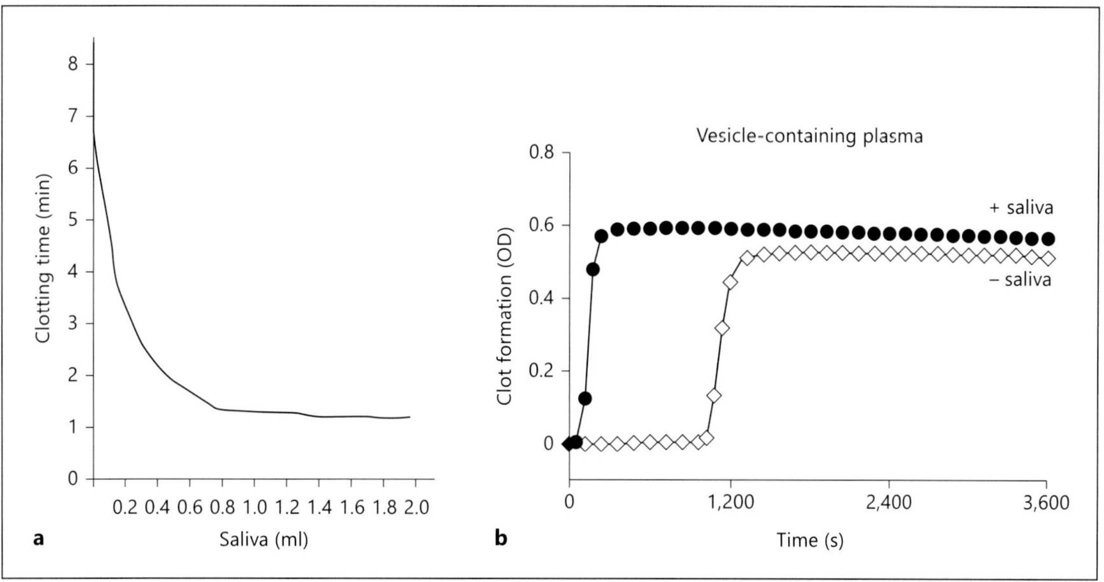

Fig. 2. a In the thirties of the previous century, it was already shown that addition of varying amounts of saliva affected the clotting time of blood [5]. **b** In a recent paper it was shown that saliva contains vesicles with coagulation factor [7].

infections by preventing microbial adhesion, killing and neutralization of microbial toxins (for a detailed treatment of antimicrobial factors in saliva, the reader is referred to the paper by van 't Hof et al. [this vol., pp. 40–51]). In addition, saliva contains several proteins and peptides that are able to stimulate directly the activity of oral epithelial cells and fibroblasts.

Growth Factors

Growth factors are signalling molecules that trigger a variety of intracellular processes, including cell division and cell migration. Epidermal growth factor (EGF) and nerve growth factor were originally discovered and characterized in saliva and salivary glands of rodents [11, 12]. Later studies revealed that these factors are also present in human saliva.

Several growth factors have been identified in human saliva. Some of these are present in concentrations that have biological activity in vitro, and could therefore potentially play a physiological role in the intra-oral wound healing. These growth factors include, among others, EGF, transforming growth factor α and vascular endothelial growth factor [13–15]. For the discovery of EGF, Stanley Cohen and Rita Levi-Montalcini were awarded the Nobel Prize in 1986. EGF is a relatively small protein of 53 amino acids, in man mainly secreted by the parotid glands. It binds to a specific receptor on the membrane of the target cell, initiating a cascade of processes. This results in migration, proliferation and differentiation of cells, activities that are important steps in the initial stages of wound healing. In rodents, salivary EGF also promotes wound healing of the skin and the stomach, in addition to its local effects in the oral cavity [16–18].

When the presence of EGF in human saliva was reported in the eighties of the previous cen-

Table 1. Concentrations of growth factors in saliva and plasma (in ng/ml) (modified from [27])

Growth factor	Human saliva	Human plasma	Murine saliva
EGF	0.9	0.2	20,000
NGF	0.9	0.1	40,000
VEGF	1.4	0.5	
FGF	<0.001	0.2	
IGF	0.4	170	75
TGF-α	5.6	0.03	560
TGF-β	0.024	2.0	
TNF-α	0.003	0.008	
Insulin	0.2	925	

tury, it was assumed that this growth factor protects the human oral epithelium as it does in rodents. However, the concentration of EGF in human saliva is considerably lower than the concentration in the saliva of rodents (table 1). In addition, a considerable part of EGF in human saliva is present in the form of an inactive precursor. Despite this, it has been shown that even in the low concentrations in which EGF is present in human saliva it has in vitro activity, suggesting that this growth factor can play a similar role in man as it does in mice and rats.

Transforming growth factor α is another growth factor present in human saliva. Its chemical structure shows a marked resemblance to that of EGF. Therefore it is not surprising that both growth factors show an almost identical range in biological activities and bind to the same receptor.

A possible role of nerve growth factor in the maintenance of oral health seems doubtful. In mice, no wound healing effects could be identified [19]. There are indications that this growth factor plays a role in the development and regeneration of nerve fibres [20].

Finally, saliva contains relatively large amounts of vascular endothelial growth factor. This growth factor is derived from the submandibular glands. It is a multifunctional protein which, among others, stimulates angiogenesis. The concentration of this growth factor is increased in the saliva of patients with periodontitis, which suggests a possible role in the healing of periodontal tissues [21].

Secretory Leucocyte Protease Inhibitor
Several proteins in saliva enhance wound healing by inhibiting the inflammatory response, for example secretory leucocyte protease inhibitor (SLPI). SLPI is an enzyme inhibitor originally isolated from human bronchial mucus. It is present in most mucosal secretions, including bronchial, nasal and cervical mucus, saliva and seminal plasma. SLPI inhibits a large number of protein-degrading enzymes, including elastase, trypsin and cathepsin. In addition, it has in vitro anti-HIV activity, anti-inflammatory activity and antimicrobial activity.

The potential role of SLPI in wound healing has been explored in animal studies. In SLPI knockout mice, missing the gene of this enzyme inhibitor, oral wound healing was delayed considerably and oral infections occurred more frequently. This was caused by an increased degradation of connective tissue by elastase. When SLPI was topically applied in SLPI-deficient mice, wound healing normalized indicating that this enzyme inhibitor is crucial for wound healing [22].

Trefoil Peptides
Another protein for which a role in oral wound healing has been suggested is trefoil factor 3 (TFF3), a member of the trefoil peptide family. Trefoil peptides are characterized by a 40-amino-acid domain containing 3 conserved disulphide bridges resulting in 3 loops. The compact form of trefoil peptides makes them remarkably resistant against proteolytic degradation. They are abundantly present on almost all mucosal surfaces of the human body, including the oral cavity, gastro-intestinal tract, gall bladder, pancreas, lungs

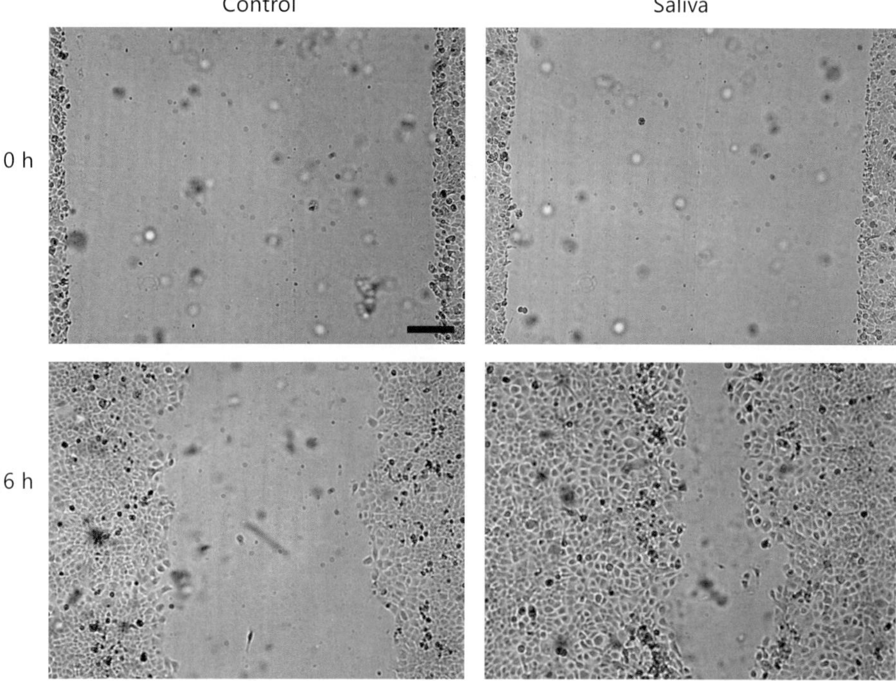

Fig. 3. Stimulating effect of parotid saliva on in vitro wound closure. In a confluent layer of epithelial cells, a scratch was made with a sterile tip, thus creating a standardized 'wound'. The width of the scratch was determined microscopically immediately after creation and 16 h after culture in medium in the absence (left panels) or presence of human parotid saliva (30% v/v, right panels).

and cervix. They improve the mechanical and chemical resistance of the mucus layer and are involved in the homeostasis and regeneration of the mucosa.

TFF3 is the only member of the trefoil peptide family present in saliva. TFF3 is secreted by the submandibular and sublingual glands. It increases wound closure in a dose-dependent manner by stimulating the migration of oral keratinocytes. TFF3 had no significant effect on the division of these cells [23].

Histatins

Histatins are a family of histidine-rich peptides that are only present in saliva of higher primates. Until now, they have not been identified in other tissues. Many functions have been attributed to histatins, including anti-inflammatory activity, detoxification and remineralization of teeth, but their antimicrobial activity has been studied most extensively.

In human saliva 26 different histatin peptides have been found, but histatins 1, 3 and 5 comprise 85% of the total histatin concentration in saliva. All histatins are derived from 2 closely related genes, HTN1 and HTN3, on chromosome 4 which encode for histatin 1 of 38 residues and histatin 3 of 32 residues. Due to proteolytic cleavage, histatin 1 gives rise to histatin 2, whereas histatin 3 gives rise to all the other isoforms.

Recently it has been discovered that salivary histatin 1 and 2 exhibit in vitro cell-stimulating properties. The discovery of the cell-stimulating properties of histatins started with the observa-

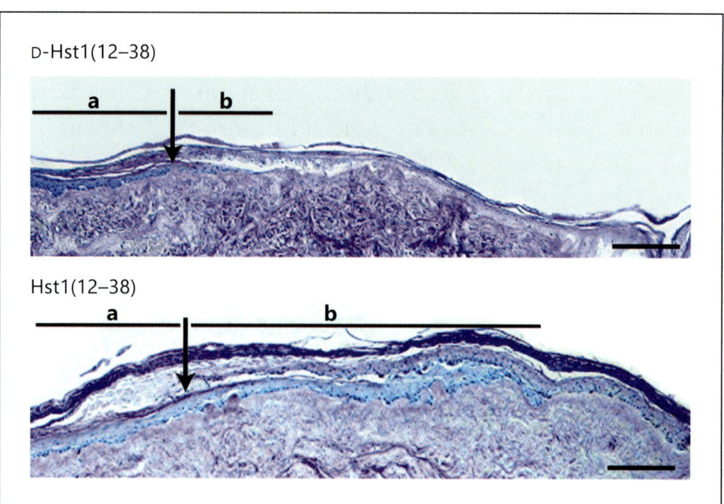

Fig. 4. Enhancement of wound healing by synthetic histatin (Hst) in a human epidermal skin equivalent that closely resembles healthy skin. Wounds were created with a glass rod cooled to –196°C. This resulted in cell death of the entire treated area of the epidermis. Immediately after creation of the wound, the epidermal equivalents were cultured for 6 days in medium in the absence (upper panel) or presence of histatin (lower panel). The arrow indicates the original border of the wound area; a = healthy epidermis; b = the length of the wound closure after 6 days.

tion that parotid saliva enhanced wound closure in a scratch assay. In this assay a scratch is made in a confluent monolayer of epithelial cells, and subsequently the closure of the gap is measured. In the presence of parotid saliva the growth of epithelial cells and closure of the scratch were enhanced (fig. 3). By fractionating parotid saliva it was found that this activity was associated specifically with histatins 1 and 2. The D-enantiomer of histatin 2, synthesized from D-amino acids, did not induce wound closure. This points to a stereospecific interaction with a receptor on the epithelial cell surface. Histatin 1 stimulates the migration of epithelial cells and fibroblasts [24], but not the proliferation. Histatin 1 also enhanced wound closure in a more complex in vitro model for wound healing, using human epidermal skin equivalents (fig. 4) [25].

The discovery of the wound-closing effects of histatins opens the way for the development of new wound healing medications. Compared to EGF, histatins are relatively simple chemical compounds. They can be synthesized chemically and can be produced in large quantities (fig. 3). The low production costs of histatins probably make them commercially more interesting than recombinant EGF and transforming growth fac-

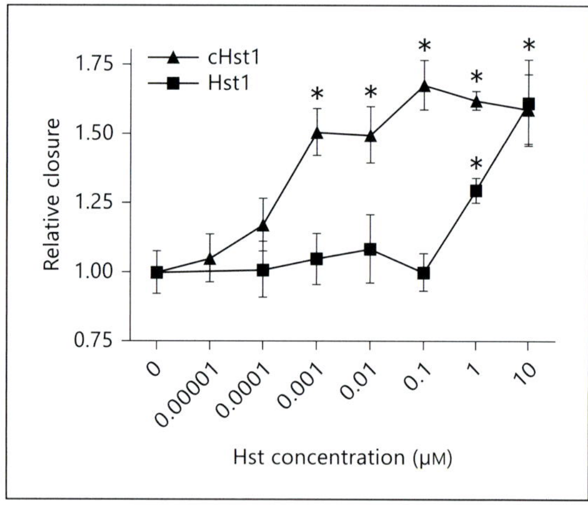

Fig. 5. Comparison of the wound-closing effects of natural linear histatin (Hst1) and synthetic cyclic histatin (cHst1), in concentrations ranging from 0.01 nM up to 10 µM. Cyclization resulted in a 1,000-fold stimulation of the molar activity. * p ≤ 0.05.

tor α, currently entering the market in small amounts.

Compared to EGF 1,000-fold higher concentrations of histatins are required for activity annihilating the advantage of lower production costs. It was hypothesized that this relative low molar activity was due to the fact that in solution histatins are present as random linear polypeptide chains, which lack a fixed 3-dimensional conformation. Indeed, cyclization of histatins, which imposes a certain structural conformation, increased the biological activity approximately 1,000-fold (fig. 5) [25]. Using sortase A for cyclization, the efficiency of histatin cyclization was optimized to approximately 90% [26].

Histatin analogues seem to have a very interesting clinical potential [27]. For example, a histatin-containing gel could be used in the oral cavity for the treatment of aphthous ulcers or to enhance the healing of mucositis. Hopefully such clinical trials will be conducted in due course.

Conclusions

It has been known for a long time that, compared to the skin, wounds in the oral cavity heal faster and with less infection and scar formation. The role of growth factors like EGF in this process is well accepted and has been investigated extensively. During the last few years it has become clear that human saliva harbours a wide variety of proteins and peptides that play a role in different phases of wound healing, including blood coagulation and cell migration. These substances may be a starting point for the development of new drugs that promote tissue healing.

References

1 Verrier L: Dog licks man. Lancet 1970;i:615.
2 Wong JW, Gallant-Behm C, Wiebe C, Mak K, Hart DA, Larjava A, Hakkinen L: Wound healing in oral mucosa results in reduced scar formation as compared with skin: evidence from the red Duroc pig model and humans. Wound Rep Regen 2009;17:717–729.
3 Engeland CG, Bosch JA, Cacioppo JT, Marucha PT: Mucosal wound healing – the roles of age and sex. Arch Surg 2006; 143:1193–1197.
4 Hunter JB: The action of saliva and gastric juice on the clotting of blood. Br J Surg 1928;16:203–207.
5 Volker JF: The effect of saliva on blood coagulation. Am J Orthodont 1939;25: 277–281.
6 Glazko AJ, Greenberg DM: The mechanism of the action of saliva in blood coagulation. Am J Physiol 1938;125:108–112.
7 Berckmans RJ, Sturk A, van Tienen LM, Schaap MCL, Nieuwland R: Cell-derived vesicles exposing coagulant tissue factor in saliva. Blood 2011;117:3172–3180.
8 Zaura E, Keijser BJF, Huse SM, Crielaard W: Defining the healthy 'core microbiome' of oral microbial communities. BMC Microbiol 2009;9:259–270.
9 Abrahamian FM, Goldstein EJC: Microbiology of animal bite wound infections. Clin Microbiol Rev 2011;24:231–246.
10 Edwards, R, Harding KG: Bacteria and wound healing. Curr Opin Infect Dis 2004;17:91–96.
11 Cohen S: Isolation of a mouse submaxillary gland protein accelerating incisor eruption and eyelid opening in the newborn animal. J Biol Chem 1962;237: 1555–1562.
12 Cohen S: Origins of growth factors: NGF and EGF. Ann NY Acad Sci 2004;1038: 98–102.
13 Royce LS, Baum BJ: Physiological levels of salivary epidermal growth-factor stimulate migration of an oral epithelial-cell line. Biochim Biophys Acta 1991; 1092:401–403.
14 Mogi M, Inagaki H, Kojima K, Minami M, Harada M: Transforming growth-factor-alpha in human submandibular-gland and saliva. Immunoassay 1995;16: 379–394.
15 Taichman NS, Cruchley AT, Fletcher LM, Hagi-Pavli EP, Paleolog E, Abrams WR, Booth V, Edwards RM, Malamud D: Vascular endothelial growth factor in normal human salivary glands and saliva. A possible role in the maintenance of mucosal homeostasis. Lab Invest 1998;78:869–875.
16 Hutson JM, Nial M, Evans D, Fowler R: Effect of salivary glands on wound contraction in mice. Nature 1979;279:793–795.
17 Olsen PS, Poulsen SS, Kirkegaard P, Nexo E: Role of submandibular saliva and epidermal growth factor in gastric cytoprotection. Gastroenterology 1984; 87:103–108.
18 Bodner L, Dayan D, Pinto Y, Hammel I: Characteristics of palatal wound healing in desalivated rats. Arch Oral Biol 1993; 38:17–21.
19 Noguchi S, Ohba Y, Oka T: Effect of salivary epidermal growth factor on wound healing of tongue in mice. Am J Physiol 1991;260:E620–E625.
20 Richardson PM, Ebendal T: Nerve growth activity in rat peripheral nerve. Brain Res 1982;246:57–64.
21 Booth V, Young S, Cruchley A, Taichman NS, Paleolog E: Vascular endothelial growth factor in human periodontal disease. J Period Res 1998;33;491–499.
22 Ashcroft GS, Lei K, Jin W, Longenecker G, Kulkarni AB, Greenwell-Wild T, Hale-Donze H, McGrady G, Song XY, Wah SM: Secretory leukocyte protease inhibitor mediates non-redundant functions necessary for normal wound healing. Nat Med 2000;6:1147–1153.

23 Storesund T, Hayashi K, Kolltveit KM, Bryne M, Schenk K: Salivary trefoil factor 3 enhances migration of oral keratinocytes. Eur J Oral Sci 2008;116:135–140.

24 Oudhoff MJ, Bolscher JGM, Nazmi K, Kalay H, van 't Hof W, Nieuw Amerongen A, Veerman ECI: Histatins are the main wound-closure stimulating factors in human saliva as identified in a cell culture assay. FASEB J 2008;22:3806–3812.

25 Oudhoff MJ, Kroeze KM, Nazmi K, van den Keijbus PAM, van 't Hof W, Fernandez-Borja M, Hordijk PL, Gibbs S, Bolscher JGM, Veerman ECI: Structure-activity analysis of histatin, a potent wound healing peptide from human saliva: cyclization potentiates molar activity 1,000-fold. FASEB J 2009;23:3928–3935.

26 Bolscher JGM, Oudhoff MJ, Nazmi K, Antos JM, Guimaraes CP, Spooner E, Haney EF, Vallejo JJG, Vogel HJ, van 't Hof W, Ploegh HL, Veerman ECI: Sortase A as a tool for high-yield histatin cyclization. FASEB J 2011;25:2650–2658.

27 Oudhoff MJ: Discovery of the Wound-Healing Capacity of Salivary Histatins; PhD thesis, Vrije Universiteit, Amsterdam, 2010.

Enno C.I. Veerman
Department of Periodontology and Oral Biochemistry
Academic Centre for Dentistry Amsterdam ACTA, Room 12-N37
Gustav Mahlerlaan 3004, NL–1081 LA Amsterdam (The Netherlands)
E-Mail e.veerman@acta.nl

Ligtenberg AJM, Veerman ECI (eds): Saliva: Secretion and Functions.
Monogr Oral Sci. Basel, Karger, 2014, vol 24, pp 61–70 (DOI: 10.1159/000358789)

Role of Saliva in Oral Food Perception

Eric Neyraud

CNRS, UMR6265 Centre des Sciences du Goût et de l'Alimentation, INRA, UMR1324 Centre des Sciences
du Goût et de l'Alimentation, et UMR Centre des Sciences du Goût et de l'Alimentation, Université de
Bourgogne, Dijon, France

Abstract

Saliva is the first fluid that comes into contact with food
during oral processing. Because saliva is the medium that
bathes the taste receptors, is the fluid through which
taste and aroma compounds are released into the oral
cavity and is mixed continuously with food during bolus
formation, it is an essential actor in oral chemosensory
perception. The complexity of saliva composition, with
compounds originating from different salivary glands,
from gingival crevicular fluid, from micro-organisms and
from food debris, together with its variable nature increases the possibilities for interactions with food compounds and for different roles in perception. These factors are increasingly being taken into account in current
research on food perception. The aim of this paper is to
review the principal roles of saliva in oral perception, with
particular focus on chemosensory perception. These include the protection of taste buds, the effects of flow
rates, salivary hormones, electrolytes and organic compounds, and finally the impact of perception on salivary
secretions.   © 2014 S. Karger AG, Basel

Saliva has numerous functions in the oral cavity
which include protecting the teeth, protecting
against micro-organisms and interactions with

food [1]. This latter function should not be underestimated because during oral processing,
saliva mixes with food and may constitute more
than 50% of bolus volume before swallowing
[2]. This suggests that when we eat, we do not
perceive the intrinsic properties of the food on
the plate but what results from interactions between the food and saliva. In addition, considerable intersubject variability is commonly observed in the context of human sensory perception, which is still difficult to explain. As there is
also marked intersubject variability in salivary
secretion [3], it was thought relevant to try to
explain sensory variability in terms of the composition of saliva, so its role in sensory perception has recently been the subject of increasing
interest.

'In-mouth' sensory perception involves per-
ceptions of texture, taste, aroma and trigeminal
sensations. The influence of saliva on texture
perception has been studied quite extensively
and will not be considered here (for reviews, see
van Aken et al. [4] and Schipper et al. [5]). Together, taste, aroma and trigeminal sensations
form the chemosensory perception most commonly known as flavour. Taste is the sensation

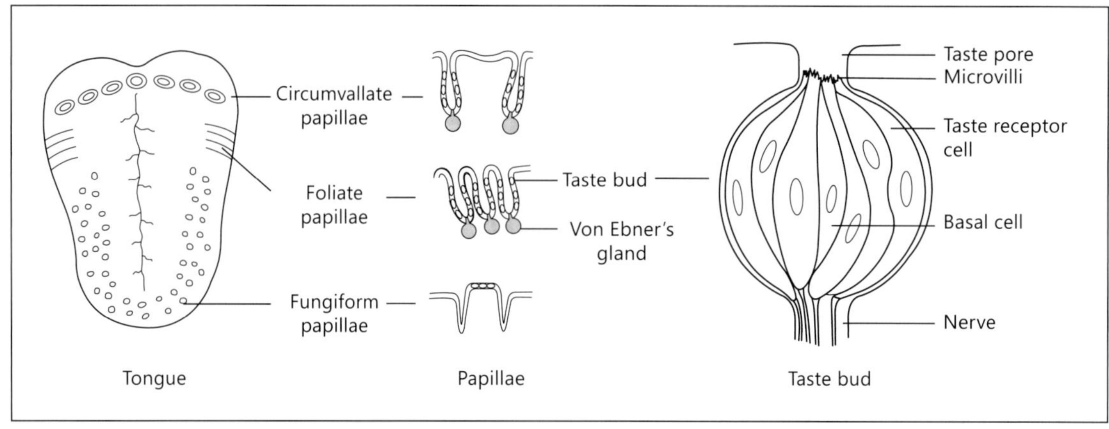

Fig. 1. Schematic representation of the tongue, of the taste papillae and of a taste bud.

produced by non-volatile substances when they react with taste bud receptors. The taste buds are mainly localized in tongue papillae, and it is worth noting that some minor salivary glands, the von Ebner glands, are localized in the cleft of circumvallate and foliate papillae (fig. 1). These glands secrete a fluid which bathes the taste receptors, strongly suggesting an important role in taste perception. The existence of 5 basic tastes – sourness, saltiness, bitterness, sweetness and umami – is commonly accepted. However, the recent discovery of fatty acid receptors in human taste buds [6] has supported the idea that fattiness may be a sixth taste. Aroma is the perception of volatile molecules by olfactory receptors in the nasal cavity when they reach it via the retronasal route. Finally, trigeminal sensations are mainly related to irritant sensations perceived by the free extremities of trigeminal nerves. A common feature of substances that can elicit these sensations is that they need to interact with saliva before reaching the receptors. It is therefore the result of these interactions that will lead to oral chemosensory perception. In this context, the aim of this paper is to review the principal mechanisms that involve saliva in human chemosensory perception.

Role of Saliva in Oral Chemosensory Perception

Protection of the Taste Buds

Saliva is the fluid which is at the interface between the taste buds and external medium. This has led to the assumption that the enzymes and hormones it contains may be implicated in the protection or modulation of taste receptor cells. Indeed, subjects who complain of hyposalivation also suffer from altered taste perception, and a direct link between salivary zinc deficiency and taste alterations was suggested [7]. This was supported by the discovery of a zinc-binding protein, gustin [8], which was subsequently identified as being carbonic anhydrase 6 [9]. Inhibition of its synthesis is associated with the development of taste bud abnormalities, and therefore it is suggested as a trophic factor affecting the taste buds [10]. In a recent study [11], it was reported that oral sensitivity to 6-n-propylthiouracil (PROP; a molecule used to determine global taste sensitivity) was inversely related to the zinc concentration in saliva, which they related to polymorphism of the gustin gene. The authors hypothesized that the PROP non-taster status is associated with a less functional form of gustin unable to bind zinc. This would explain why zinc supplementation is not always efficient in patients with taste dysfunctions [12]. Other mechanisms lie at the origin of hypogeusia. For instance, lower salivary cAMP and cGMP concentrations in patients suffering from hypogeusia have been reported [13] suggesting a possible role for salivary cAMP as a growth factor in taste buds [14].

Salivary Hormones Involved in Chemosensory Perception

Numerous circulating hormones are present in saliva, and a variety of proteins that can be detected in saliva appear to originate from blood [15]. In a recent article, Zolotukhin [16] reviewed the different metabolic hormones present in saliva and their functions. Quite a few play a direct role in taste perception (often by modulating taste receptors); although most of these studies were performed in animal models, some of these hormones were also found in human saliva, such as glucagon [17], insulin [18], leptin [19, 20], oxytocin [21] and ghrelin [20]. These hormones have been reported to modulate all basic taste modalities. $Scg5^{-/-}$ mice, which lack mature glucagon, display a significantly reduced responsiveness to sucrose when compared to wild-type animals through a local action, as glucagon and its receptor are co-expressed in a subset of mouse taste receptor cells [22]. Leptin appeared to modulate sweet taste because circadian variations in this hormone seem to be linked to variations in sweetness sensitivity [16]. The administration of leptin in lean mice suppressed the responses of peripheral taste nerves to sweet substances, suggesting a sweet-sensing suppressor effect [23]. Ghrelin receptor null mice exhibited a significantly reduced taste response to sour (citric acid) and salty (NaCl) tastants [24]. Finally, salivary insulin enhanced the salty taste modality in mice [25]. Other hormones also seem to be implicated in modulating taste perception, but their presence in saliva has not yet been reported, and they probably exert an autocrine action. That is the case of glucagon-like peptide 1, which is involved in sweet, sour and umami tastes [26], or vasoactive intestinal peptide for sweet, bitter and sour tastes [27].

Role of Flow Rates and Electrolytes

As well as its roles in protecting or modulating taste receptors, saliva is also involved in perception through its interactions with sensory stimuli. This is firstly due to the volume of this fluid, which plays a determinant role in taste perception because in order to reach receptors taste compounds must be diluted in liquid. The volume of saliva will therefore influence the concentration of taste compounds at the level of receptors, and some studies have reported better correlations between taste perception and the concentrations of taste compounds in saliva than with the initial concentration in the food matrix [28, 29]. Some authors have also tried to demonstrate a direct effect of the salivary flow rate on perception by comparing taste or aroma perception in subjects varying in their salivary flow rate, but the findings were contradictory. A longer persistence of bitterness and astringency has been reported in low-flow subjects [30] or during wine consumption [31]. On the other hand, no effects of flow rate were reported with sour, sweet or fruity solutions [32]. During a recent study [33], the salivary flow rates of some individuals were modified by adding artificial saliva near the outflow of the parotid ducts using a modified Lashley cup. The authors reported a reduction in the perception of citric acid and sodium chloride solutions which was probably due to diluting effects, but not for bitterness or sweetness, suggesting the involvement of other mechanisms. It is therefore complicated to conclude as to a general effect of salivary flow rate on perception since the mechanisms involved are dependent on the type of food matrix studied (solid, liquid, with different compositions and properties), the nature of the molecule being perceived and potential interactions with organic or inorganic salivary compounds.

Inorganic compounds in saliva play an important role in the taste perception of ions through various mechanisms, an important one being adaptation. Saliva contains different minerals, including sodium. Consequently, the taste buds are adapted to salivary Na^+ concentrations and to be perceived, the Na^+ concentration arising from a stimulus must be higher than that in secreted saliva. Similarly, the NaCl detection threshold is lower after the mouth has been rinsed with deionized water than when it is adapted to saliva [34]. In addition, because the Na^+ concentration in saliva increases in line with the flow rate, because of the reabsorption mechanisms at work in salivary gland ducts [35], Na^+ sensitivity is lower at high flow rates such as those which occur during chewing [36]. Although the case of NaCl has been the most widely reported, adaptation effects can also occur with other substances. For instance, subjects with low monosodium glutamate concentrations in their saliva find this taste (umami) more unpleasant [37]. A second mechanism concerns the perception of sourness. In the mouth, this perception increases when the pH falls, but it is also dependent on titrable acidity [38] because at an equal pH, weak acids have been shown to elicit a stronger response than HCl during psychophysical experiments in humans [39]. Indeed, one of the main functions of saliva is to regulate the oral pH, especially by secreting bicarbonate ions. Saliva thus modulates this perception through its buffering capacity [40, 41]. Furthermore, the buffering capacity of saliva increases in line with flow rate [42], leading to a higher buffering capacity during stimulation by acids, which are well known to be the most effective stimuli of salivary secretions [43, 44].

Interactions between Saliva Organic Compounds and Taste and Aroma Molecules
Saliva contains a broad diversity of organic molecules including proteins – of which more than 1,000 have now been identified [45] –, metabolites and nucleic acids. Research to date has determined the roles of certain proteins and their mechanisms of action, which can be divided into three main categories, enzymatic conversion, binding and transport.

Enzymatic Conversion
About 30 different enzymatic activities have been reported in human saliva [46]. Saliva contains enzymes that can lyse the principal macronutrients, which include complex carbohydrates with amylase [47, 48], proteins with protease [49] and lipids with lipase [50–52]. The contribution of these enzymes to food digestion appears to be small, in view of the weak activities reported [51], with the possible exception of amylase. Their actions may be related to perception, having a role in the conversion of food macronutrients into molecules that can be perceived by the taste or olfactory systems. Indeed, a correlation has been seen between the lipase activity measured in resting whole saliva and the intensity of perception of a triglyceride solution [51]. The hypothesis is that the hydrolysis of triglycerides by lipase may cause the release of free fatty acids which can then be perceived by the taste system. Following that study, it was also found that lipase activity was positively correlated with the oleic acid taste threshold [53]. A direct link between proteolytic enzymes in saliva and perception has not yet been established. However, recent work on bitterness perception [54] showed that a greater sensitivity to caffeine perception was associated with a lower level of cystatin SN, a protease inhibitor, and a higher representation of protein fragments which probably resulted from a high degree of proteolysis. This suggests that proteolysis may be a perireceptor event that plays a role in bitterness perception. The implication of α-amylase in taste perception has also been shown indirectly. This is likely due to the fact that the final product of starch hydrolysis by α-amylase is maltose, which does not have a strong sweet taste. However, an association

between α-amylase activity and a reduction in saltiness perception in NaCl-containing starch-thickened foods has been shown [55]. This unexpected correlation was explained by a disruption of the starch structure by the enzyme and a reduction in the NaCl dilution in the saliva.

The involvement of other salivary enzymes in perception has also been demonstrated, such as carbonic anhydrase in the perception of carbonated beverages [56], and their role in retronasal olfaction has been strongly suggested. For instance, in white wine treated with human saliva, levels of esters and fusel alcohols (responsible for fruity and fusel oil odours) were reduced by 32 and 80%; by contrast, concentrations of 2-phenylethanol and furfural (responsible for rose and toasted almond notes) increased by 27 and 155%, respectively [57]. Such conversions can be attributed, at least in part, to enzymatic conversion [58, 59].

Binding
The binding of taste or flavour compounds by salivary components is thought to play an important role in their perception by modulating the concentration of free compounds available to receptors. This was discussed in a recent study [60] on aroma release during the in vivo consumption of a sweet mint tablet, when a positive correlation was found between tablet degradation and menthone release, and a negative one between tablet degradation and salivary protein levels. The authors suggest interactions between aroma compounds and salivary proteins. However, the influence of saliva on the partitioning of aroma compounds has mainly been shown in vitro using 'artificial' saliva. This type of solution usually consists of a mixture of the principal inorganic molecules found in saliva with mucin and/or amylase [61, 62]; it is not yet known whether these results can be extrapolated to the properties of human saliva.

The most widely studied binding properties of salivary proteins concern proline-rich proteins (PRPs) and their role in the perception of astringency. PRPs are a family of intrinsically disordered proteins with considerable affinity for tannins. Their binding property is a defence mechanism designed to neutralize tannins so as to prevent their harmful effects on digestion [63]. As well as this physiological role, these interactions are probably involved in the perception of astringency. Indeed, it appears that PRPs scavenge tannins up to a certain level of concentration, but when their scavenging capacity is exceeded, tannins may interact with other salivary proteins, causing changes to the lubricating properties of saliva and the perception of roughness [64]. To support this hypothesis, recent findings showed that a lower protein concentration in saliva was associated with a higher perception of astringency [65]. Furthermore, subjects who can rapidly restore their salivary protein levels after stimulation are less sensitive to astringency because these levels are sufficient to successfully intercept and inactivate dietary tannins [66].

PRPs may also be implicated in taste perception. Basal levels of two of them (ll-2 and Ps-1) were found to be significantly higher in PROP supertaster compared to PROP non-taster unstimulated saliva; the same study also reported that PROP stimulation elicited a rapid rise in the levels of these proteins, but only in PROP supertaster saliva [67]. During a subsequent study performed by the same group [68], the authors found that in individuals who lacked Ps-1 in their saliva, Ps-1 supplementation enhanced their responsiveness to a PROP bitter taste. The authors suggested that Ps-1, and particularly the L-arginine present in this protein, may facilitate PROP binding to the receptor.

Transport
Because taste compounds need to be dissolved in saliva to reach taste receptors, an important question which remains concerns hydrophobic compounds. This issue has become relevant since the

existence of a sixth taste was suggested with the discovery of fatty acid taste receptors in humans (for a review, see Mattes [6]). In order to reach these receptors, the presence of molecules able to solubilize hydrophobic compounds in saliva is required. In 1993 [69], a protein was identified in human von Ebner salivary glands and called the von Ebner gland protein. This belongs to the lipocalin family and has strong homologies with olfactory binding proteins which are able to transport volatile aroma compounds to the olfactory receptors via the olfactory mucus. Von Ebner gland protein (also called LCN1) is capable of binding hydrophobic molecules such as fatty acids [70]. Its specific secretion by the lingual von Ebner gland close to the taste buds would thus create a salivary medium with a specific composition at that site, capable of carrying fatty acids to the taste receptors (fig. 1).

Impact of Perception on Saliva

In 1916, Lashley [71] reported that the secretion of saliva was induced by taste stimulation through a parotid salivary reflex. Many subsequent studies focused on the effects of different taste compounds on salivary secretion and reported the effects of different tastes (sour, salty, sweet, bitter and umami); stimulation induced by acids was the most effective and the flow rate increased in line with the concentration of the stimulus [72–75]. Other sensory stimulations also elicit salivary secretions; this is the case of olfaction, even if the effects reported are contradictory [76] and seem to depend on the type of olfactant. The mechanisms triggering these secretions have not yet been clearly identified but it appears that they may result from association effects because a lemon flavour is one of the best stimuli [77, 78]. It is interesting to note that as a counterbalance to taste stimulation, the submandibular and sublingual glands contribute most to these secretions [79].

One remarkable point is that the composition of saliva appears to differ as a function of the type of stimulation [80]. Whole saliva collected after stimulation with an acidic beverage has greater elasticity when compared to stimulation by water or chewing [81, 82]. The authors suggested that this is probably a defence mechanism designed to protect the teeth from acid erosion in response to acidic beverages. Similarly, Neyraud et al. [83] reported that the protein composition (proteome) of whole saliva differed after stimulation with various tastants, the most aversive ones inducing the most marked modifications following this order: sour, bitter, umami and sweet. These authors made the assumption that the proteins overexpressed under these conditions form part of an anti-inflammatory response targeting the effects of harmful compounds. Modifications of the whole-saliva proteome following stimulation with tastants were also reported recently by other groups [84–86]. Insofar as whole saliva was analysed during these studies, and Neyraud et al. [83] could not determine any changes to the parotid saliva proteome after stimulation with bitter compounds, the origins of these modifications remain unknown. These modifications may therefore be due to differences in the proportion between secretions by different salivary glands, secretions by the mucosa or those of gingival crevicular fluid. However, some studies have shown that modifications of saliva composition following a sensory stimulation may be gland specific. Indeed, after stimulation with various tastants, the protein concentration in saliva collected from parotid glands was significantly higher after stimulation with citric acid than with sucrose, monosodium glutamate and $MgSO_4$, at equal flow rates [75]. Similarly, an increase in α-amylase activity has been shown in parotid saliva after an intake of sucrose solution but not after sucrose sham feeding [87], and the lipase activity and anti-oxidant capacity measured in whole saliva differed after stimulation with oleic acid compared to a control in subjects who were hypersensitive to the taste of this fatty acid [88]. So even though the underlying mechanisms are not known, these observations open new perspectives concerning how the flow rate and composition of saliva may be modified in order to respond adequately to different stimulations or the food being consumed.

Concluding Remarks

Saliva is central to oral chemosensory perception in many ways, which includes protection of the taste buds, the effects of flow rates and electrolyte levels and the roles of different proteins. The varying relationships observed between saliva composition and sensory perception show that saliva is key to explaining the differences in food perception, acceptability and behaviour seen in the population. The development of 'omics' approaches – and particularly that of proteomics – has enabled the proposal of new hypotheses. The role of the salivary metabolome has been little investigated to date, but this field appears promising because some relationships have already been observed between it and diet [89] or stimulation [90].

Saliva also has the remarkable property of being modified by sensory stimulation and perhaps of adapting to the type of stimulus. To date, most results have been of an observational nature, and the underlying mechanisms remain unknown, thus opening the way to fascinating prospects for research.

Acknowledgements

Dr. Martine Morzel and Dr. Francis Canon are thanked for their valuable advice and Arnaud Lafaye for having kindly drawn the figure.

References

1 Amerongen AVN, Veerman ECI: Saliva – the defender of the oral cavity. Oral Dis 2002;8:12–22.

2 Repoux M, Laboure H, Courcoux P, Andriot I, Semon E, Yven C, Feron G, Guichard E: Combined effect of cheese characteristics and food oral processing on in vivo aroma release. Flavour Frag J 2012;27:414–423.

3 Quintana M, Palicki O, Lucchi G, Ducoroy P, Chambon C, Salles C, Morzel M: Inter-individual variability of protein patterns in saliva of healthy adults. J Proteomics 2009;72:822–830.

4 Van Aken GA, Vingerhoeds MH, de Hoog EHA: Food colloids under oral conditions. Curr Opin Colloid Interface Sci 2007;12:251–262.

5 Schipper RG, Silletti E, Vinyerhoeds MH: Saliva as research material: biochemical, physicochemical and practical aspects. Arch Oral Biol 2007;52:1114–1135.

6 Mattes RD: Is there a fatty acid taste? Annu Rev Nutr 2009;29:305–327.

7 Henkin RI, Bradley DF: Hypogeusia corrected by Ni^{++} and Zn^{++}. Life Sci II 1970;9:701–709.

8 Henkin RI, Lippoldt RE, Bilstad J, Edelhoch H: Zinc protein isolated from human parotid saliva. Proc Natl Acad Sci USA 1975;72:488–492.

9 Thatcher BJ, Doherty AE, Orvisky E, Martin BM, Henkin RI: Gustin from human parotid saliva is carbonic anhydrase VI. Biochem Biophys Res Commun 1998;250:635–641.

10 Henkin RI, Martin BM, Agarwal RP: Decreased parotid saliva gustin/carbonic anhydrase VI secretion: an enzyme disorder manifested by gustatory and olfactory dysfunction. Am J Med Sci 1999;318:380–391.

11 Padiglia A, Zonza A, Atzori E, Chillotti C, Calo C, Tepper BJ, Barbarossa IT: Sensitivity to 6-n-propylthiouracil is associated with gustin (carbonic anhydrase VI) gene polymorphism, salivary zinc, and body mass index in humans. Am J Clin Nutr 2010;92:539–545.

12 Matsuo R: Role of saliva in the maintenance of taste sensitivity. Crit Rev Oral Biol Med 2000;11:216–229.

13 Henkin RI, Velicu I, Papathanassiu A: cAMP and cGMP in human parotid saliva: relationships to taste and smell dysfunction, gender, and age. Am J Med Sci 2007;334:431–440.

14 Henkin RI, Velicu I: Differences between and within human parotid saliva and nasal mucus cAMP and cGMP in normal subjects and in patients with taste and smell dysfunction. J Oral Pathol Med 2011;40:504–509.

15 Miller CS, Foley JD, Bailey AL, Campell CL, Humphries RL, Christodoulides N, Floriano PN, Simmons G, Bhagwandin B, Jacobson JW, Redding SW, Ebersole JL, McDevitt JT: Current developments in salivary diagnostics. Biomark Med 2010;4:171–189.

16 Zolotukhin S: Metabolic hormones in saliva: origins and functions. Oral Dis 2013;19:219–229.

17 Perez-Castillo A, Blazquez E: Synthesis and release of glucagon by human salivary glands. Diabetologia 1980;19:123–129.

18 Fekete Z, Korec R, Feketeova E, Murty VLN, Piotrowski J, Slomiany A, Slomiany BL: Salivary and plasma-insulin levels in man. Biochem Mol Biol Int 1993;30:623–629.

19 De Matteis R, Puxeddu R, Riva A, Cinti S: Intralobular ducts of human major salivary glands contain leptin and its receptor. J Anat 2002;201:363–370.

20 Aydin S, Halifeoglu I, Ozercan IH, Erman F, Kilic N, Ilhan N, Ozkan Y, Akpolat N, Sert L, Caylak E: A comparison of leptin and ghrelin levels in plasma and saliva of young healthy subjects. Peptides 2005;26:647–652.

21 White-Traut R, Watanabe K, Pournajafi-Nazarloo H, Schwertz D, Bell A, Carter CS: Detection of salivary oxytocin levels in lactating women. Dev Psychobiol 2009;51:367–373.

22 Elson AET, Dotson CD, Egan JM, Munger SD: Glucagon signaling modulates sweet taste responsiveness. FASEB J 2010;24:3960–3969.

23 Shigemura N, Ohta R, Kusakabe Y, Miura H, Hino A, Koyano K, Nakashima K, Ninomiya Y: Leptin modulates behavioral responses to sweet substances by influencing peripheral taste structures. Endocrinology 2004;145:839–847.

24 Shin YK, Martin B, Kim W, White CM, Ji SG, Sun YX, Smith RG, Sevigny J, Tschop MH, Maudsley S, Egan JM: Ghrelin is produced in taste cells and ghrelin receptor null mice show reduced taste responsivity to salty (NaCl) and sour (citric acid) tastants. PLoS One 2010;5:e12729.

25 Baquero AF, Gilbertson TA: Insulin activates epithelial sodium channel (ENaC) via phosphoinositide 3-kinase in mammalian taste receptor cells. Am J Physiol Cell Physiol 2011;300:C860–C871.

26 Martin B, Dotson CD, Shin YK, Ji SG, Drucker DJ, Maudsley S, Munger SD: Modulation of taste sensitivity by GLP-1 signaling in taste buds; in Finger TE (ed): International Symposium on Olfaction and Taste. Oxford, Blackwell Publishing, 2009, pp 98–101.

27 Martin B, Shin YK, White CM, Ji S, Kim W, Carlson OD, Napora JK, Chadwick W, Chapter M, Waschek JA, Mattson MP, Maudsley S, Egan JM: Vasoactive intestinal peptide-null mice demonstrate enhanced sweet taste preference, dysglycemia, and reduced taste bud leptin receptor expression. Diabetes 2010;59:1143–1152.

28 Neyraud E, Prinz J, Dransfield E: NaCl and sugar release, salivation and taste during mastication of salted chewing gum. Physiol Behav 2003;79:731–737.

29 Neyraud E, Dransfield E: Relating ionisation of calcium chloride in saliva to bitterness perception. Physiol Behav 2004;81:505–510.

30 Fischer U, Boulton RB, Noble AC: Physiological factors contributing to the variability of sensory assessments: relationship between salivary flow rate and temporal perception of gustatory stimuli. Food Qual Prefer 1994;5:55–64.

31 Noble AC: Application of time-intensity procedures for the evaluation of taste and mouthfeel. Am J Enol Vitic 1995;46:128–133.

32 Bonnans SR, Noble AC: Interaction of salivary flow with temporal perception of sweetness, sourness, and fruitiness. Physiol Behav 1995;57:569–574.

33 Heinzerling CI, Stieger M, Bult JHF, Smit G: Individually modified saliva delivery changes the perceived intensity of saltiness and sourness. Chemosens Percept 2011;4:145–153.

34 Bartoshuk LM: Psychophysics of taste. Am J Clin Nut 1978;31:1068–1077.

35 Poulsen JH: Secretion of electrolytes and water by salivary glands; in Garret JR, Ekström J, Anderson LC (eds): Glandular Mechanisms of Salivary Secretion. Basel, Karger, 1998, pp 55–72.

36 Delwiche J, Omahony M: Changes in secreted salivary sodium are sufficient to alter salt taste sensitivity: use of signal detection measures with continuous monitoring of the oral environment. Physiol Behav 1996;59:605–611.

37 Scinska-Bienkowska A, Wrobel E, Turzynska D, Bidzinski A, Jezewska E, Sienkiewicz-Jarosz H, Golembiowska K, Kostowski W, Kukwa A, Plaznik A, Bienkowski P: Glutamate concentration in whole saliva and taste responses to monosodium glutamate in humans. Nutr Neurosci 2006;9:25–31.

38 Norris MB, Noble AC, Pangborn RM: Human-saliva and taste responses to acids varying in anions, titrable acidity, and pH. Physiol Behav 1984;32:237–244.

39 Ganzevles PGJ, Kroeze JHA: The sour taste of acids – the hydrogen-ion and the undissociated acid as sour agents. Chem Senses 1987;12:563–576.

40 Christensen CM, Brand JG, Malamud D: Salivary changes in solution pH – a source of individual-differences in sour taste perception. Physiol Behav 1987;40:221–227.

41 Lugaz O, Pillias AM, Boireau-Ducept N, Faurion A: Time-intensity evaluation of acid taste in subjects with saliva high flow and low flow rates for acids of various chemical properties. Chem Senses 2005;30:89–103.

42 Neyraud E, Bult JHF, Dransfield E: Continuous analysis of parotid saliva during resting and short-duration simulated chewing. Arch Oral Biol 2009;54:449–456.

43 Chauncey HH, Feller RP, Shannon IL: Effect of acid solutions on human gustatory chemoreceptors as determined by parotid gland secretion rate. Proc Soc Exp Biol Med 1963;112:917–923.

44 Feller RP, Sharon IM, Chauncey HH, Shannon IL: Gustatory perception of sour, sweet, and salt mixtures using parotid gland flow rate. J Appl Physiol 1965;20:1341–1344.

45 Huq NL, Cross KJ, Ung M, Myroforidis H, Veith PD, Chen D, Stanton D, He H, Ward BR, Reynolds EC: A review of the salivary proteome and peptidome and saliva-derived peptide therapeutics. Int J Pept Res Ther 2007;13:547–564.

46 Salles C, Chagnon MC, Feron G, Guichard E, Laboure H, Morzel M, Semon E, Tarrega A, Yven C: In-mouth mechanisms leading to flavor release and perception. Crit Rev Food Sci Nutr 2011;51:67–90.

47 De Wijk RA, Prinz JF, Engelen L, Weenen H: The role of alpha-amylase in the perception of oral texture and flavour in custards. Physiol Behav 2004;83:81–91.

48 Engelen L, van den Keybus PAM, de Wijk RA, Veerman ECI, Amerongen AVN, Bosman F, Prinz JF, van der Bilt A: The effect of saliva composition on texture perception of semi-solids. Arch Oral Biol 2007;52:518–525.

49 Helmerhorst EJ: Whole saliva proteolysis – wealth of information for diagnostic exploitation; in Malamud D, Niedbala RS (eds): Oral-Based Diagnostics. Oxford, Blackwell Publishing, 2007, pp 454–460.

50 Drago SR, Panouille M, Saint-Eve A, Neyraud E, Feron G, Souchon I: Relationships between saliva and food bolus properties from model dairy products. Food Hydrocolloids 2011;25:659–667.

51 Neyraud E, Palicki O, Schwartz C, Nicklaus S, Feron G: Variability of human saliva composition: possible relationships with fat perception and liking. Arch Oral Biol 2012;57:556–566.

52 Stewart JE, Feinle-Bisset C, Golding M, Delahunty C, Clifton PM, Keast RSJ: Oral sensitivity to fatty acids, food consumption and BMI in human subjects. Br J Nutr 2010;104:145–152.

53 Poette J, Mekoue J, Neyraud E, Berdeaux O, Renault A, Guichard E, Genot C, Feron G: Fat sensitivity in human: oleic acid detection thresholds in model emulsion is linked to saliva composition and oral volume. Flavour Frag J 2014;29:39–49.

54 Dsamou M, Palicki O, Septier C, Chabanet C, Lucchi G, Ducoroy P, Chagnon MC, Morzel M: Salivary protein profiles and sensitivity to the bitter taste of caffeine. Chem Senses 2012;37:87–95.

55 Ferry ALS, Mitchell JR, Hort J, Hill SE, Taylor AJ, Lagarrigue S, Valles-Pamies B: In-mouth amylase activity can reduce perception of saltiness in starch-thickened foods. J Agric Food Chem 2006;54: 8869–8873.

56 Dessirier JM, Simons CT, Carstens MI, O'Mahony M, Carstens E: Psychophysical and neurobiological evidence that the oral sensation elicited by carbonated water is of chemogenic origin. Chem Senses 2000;25:277–284.

57 Genovese A, Piombino P, Gambuti A, Moio L: Simulation of retronasal aroma of white and red wine in a model mouth system. Investigating the influence of saliva on volatile compound concentrations. Food Chem 2009;114:100–107.

58 Buettner A: Influence of human saliva on odorant concentrations. 2. Aldehydes, alcohols, 3-alkyl-2-methoxypyrazines, methoxyphenols, and 3-hydroxy-4,5-dimethyl-2(^{5}H)-furanone. J Agric Food Chem 2002;50:7105–7110.

59 Buettner A: Influence of human salivary enzymes on odorant concentration changes occurring in vivo. 1. Esters and thiols. J Agric Food Chem 2002;50: 3283–3289.

60 Repoux M, Semon E, Feron G, Guichard E, Laboure H: Inter-individual variability in aroma release during sweet mint consumption. Flavour Frag J 2012;27: 40–46.

61 Friel EN, Taylor AJ: Effect of salivary components on volatile partitioning from solutions. J Agric Food Chem 2001;49:3898–3905.

62 Van Ruth SM, Grossmann I, Geary M, Delahunty CM: Interactions between artificial saliva and 20 aroma compounds in water and oil model systems. J Agric Food Chem 2001;49:2409–2413.

63 Canon F, Ballivian R, Chirot F, Antoine R, Sarni-Manchado P, Lemoine J, Dugourd P: Folding of a salivary intrinsically disordered protein upon binding to tannins. J Am Chem Soc 2011;133: 7847–7852.

64 Canon F, Pate F, Cheynier V, Sarni-Manchado P, Giuliani A, Perez J, Durand D, Li J, Cabane B: Aggregation of the salivary proline-rich protein IB5 in the presence of the tannin EgCG. Langmuir 2013;29:1926–1937.

65 Nayak A, Carpenter GH: A physiological model of tea-induced astringency. Physiol Behav 2008;95:290–294.

66 Dinnella C, Recchia A, Fia G, Bertuccioli M, Monteleone E: Saliva characteristics and individual sensitivity to phenolic astringent stimuli. Chem Senses 2009; 34:295–304.

67 Cabras T, Melis M, Castagnola M, Padiglia A, Tepper BJ, Messana I, Barbarossa IT: Responsiveness to 6-n-propylthiouracil (PROP) is associated with salivary levels of two specific basic proline-rich proteins in humans. PLoS One 2012; 7:e30962.

68 Melis M, Aragoni MC, Arca M, Cabras T, Caltagirone C, Castagnola M, Crnjar R, Messana I, Tepper BJ, Barbarossa IT: Marked increase in PROP taste responsiveness following oral supplementation with selected salivary proteins or their related free amino acids. PLoS One 2013;8:e59810.

69 Blaker M, Kock K, Ahlers C, Buck F, Schmale H: Molecular-cloning of human von Ebner's gland protein, a member of the lipocalin superfamily highly expressed in lingual salivary-glands. Biochim Biophys Acta 1993;1172:131–137.

70 Glasgow BJ, Abduragimov AR, Farahbakhsh ZT, Faull KF, Hubbell WL: Tear lipocalins bind a broad array of lipid ligands. Curr Eye Res 1995;14:363–372.

71 Lashley KS: Reflex secretion of the human parotid gland. J Exp Psychol 1916; 1:461–493.

72 Froehlich DA, Pangborn RM, Whitaker JR: The effect of oral-stimulation on human-parotid salivary flow-rate and alpha-amylase secretion. Physiol Behav 1987;41:209–217.

73 Watanabe S, Dawes C: The effects of different foods and concentrations of citric-acid on the flow-rate of whole saliva in man. Arch Oral Biol 1988;33:1–5.

74 Hodson NA, Linden RWA: The effect of monosodium glutamate on parotid salivary flow in comparison to the response to representatives of the other four basic tastes. Physiol Behav 2006;89:711–717.

75 Neyraud E, Heinzerling CI, Bult JHF, Mesmin C, Dransfield E: Effects of different tastants on parotid saliva flow and composition. Chemosens Percept 2009;2:108–116.

76 Spence C: Mouth-watering: the influence of environmental and cognitive factors on salivation and gustatory/flavor perception. J Texture Stud 2011;42: 157–171.

77 Pangborn RM, Witherly SA, Jones F: Parotid and whole-mouth secretion in response to viewing, handling, and sniffing food. Perception 1979;8:339–346.

78 Pangborn RM: Parotid flow stimulated by the sight, feel and odor of lemon. Percept Mot Skills 1968;27:1340–1342.

79 Ilangakoon Y, Carpenter GH: Is the mouthwatering sensation a true salivary reflex? J Texture Stud 2011;42:212–216.

80 Dawes C: Stimulus effects on protein and electrolyte concentrations in parotid-saliva. J Physiol 1984;346:579–588.

81 Stokes JR, Davies GA: Viscoelasticity of human whole saliva collected after acid and mechanical stimulation. Biorheology 2007;44:141–160.

82 Davies GA, Wantling E, Stokes JR. The influence of beverages on the stimulation and viscoelasticity of saliva: relationship to mouthfeel? Food Hydrocolloids 2009;23:2261–2269.

83 Neyraud E, Sayd T, Morzel M, Dransfield E: Proteomic analysis of human whole and parotid salivas following stimulation by different tastes. J Proteome Res 2006;5:2474–2480.

84 Quintana M, Palicki O, Lucchi G, Ducoroy P, Chambon C, Salles C, Morzel M: Short-term modification of human salivary proteome induced by two bitter tastants, urea and quinine. Chemosens Percept 2009;2:133–142.

85 Lorenz K, Bader M, Klaus A, Weiss W, Gorg A, Hofmann T: Orosensory stimulation effects on human saliva proteome. J Agric Food Chem 2011;59:10219–10231.

86 Silletti E, Bult JHF, Stieger M: Effect of NaCl and sucrose tastants on protein composition of oral fluid analysed by SELDI-TOF-MS. Arch Oral Biol 2012;57:1200–1210.

87 Harthoorn LF, Brattinga C, Van Kekem K, Neyraud E, Dransfield E: Effects of sucrose on salivary flow and composition: differences between real and sham intake. Int J Food Sci Nutr 2009;60:637–646.

88 Mounayar R, Septier C, Chabanet C, Feron G, Neyraud E: Oral fat sensitivity in humans: links to saliva composition before and after stimulation by oleic acid. Chemosens Percept 2013;6:118–126.

89 Sugimoto M, Saruta J, Matsuki C, To M, Onuma H, Kaneko M, Soga T, Tomita M, Tsukinoki K: Physiological and environmental parameters associated with mass spectrometry-based salivary metabolomic profiles. Metabolomics 2013;9:454–463.

90 Neyraud E, Tremblay-Franco M, Gregoire S, Berdeaux O, Canlet C: Relationships between the metabolome and the fatty acid composition of human saliva: effects of stimulation. Metabolomics 2013;9:213–222.

Eric Neyraud
INRA, Centre des Sciences du Goût et de l'Alimentation
17, rue Sully, FR–21065 Dijon Cedex (France)
E-Mail eric.neyraud@dijon.inra.fr

Ligtenberg AJM, Veerman ECI (eds): Saliva: Secretion and Functions.
Monogr Oral Sci. Basel, Karger, 2014, vol 24, pp 71–87 (DOI: 10.1159/000358790)

Lubrication

Gleb E. Yakubov

School of Chemical Engineering and Australian Research Council Centre of Excellence in Plant Cell Walls,
University of Queensland, St. Lucia, Qld., Australia

Abstract

Saliva is capable of decreasing friction force by at least 2 orders of magnitude when in between hydrophobic surfaces. This ability to lubricate is key to oral health, food processing and taste perception. In this paper different mechanisms of saliva lubrication are reviewed, and their interconnection is demonstrated using a simple physical framework. The current understanding of the roles of the molecular structure and physicochemical properties of major salivary proteins and protein complexes on lubrication is summarised and critically evaluated.

© 2014 S. Karger AG, Basel

A Brief Introduction to the Theory of Friction

Even the slightest sweep of the tongue over the back of our teeth results in a complex path of mechanical motions and energy conversions that give rise to friction. From the friction perspective, the oral environment is very complex due to a diverse range of rubbing contacts that are formed between soft tissues such as the tongue and the hard palate or hard surfaces like teeth [1]. The loss of lubrication can impede proper speech, mastication and swallowing, as well as underlie excessive friction and tooth wear. A temporal loss of salivary lubrication may also be inflicted by the consumption of foods and beverages rich in polyphenols, like red wine and tea, in which the loss of salivary lubrication is a key factor responsible for the perception of the mouth feeling called astringency [2].

The definition of friction is given as the force acting to resist the motion between surfaces that are either in a static or rubbing contact. When molecules from opposite surfaces collide, the linear momentum, $p = mv$ (where m is the mass and v is velocity), is transferred between them, and the kinetic energy ($1/2mv^2$) is transformed (or using the physicists' terminology 'dissipated') into heat, sound, mechanical deformation and even emission of light. Since the energy dissipation process is dynamic, friction between moving surfaces is called dynamic friction. This dynamic friction is responsible for the heating of rubbing surfaces, squeaky wheels, and ultimately leads to material wear such as abrasive and attritive tooth wear. In addition to the dynamic friction, an additional barrier that has to be overcome is the adhesive interactions between surfaces, e.g. attractive van der Waals or hydrophobic forces. This effect, called static friction, dominates when non-moving surfaces stick to each other, thereby

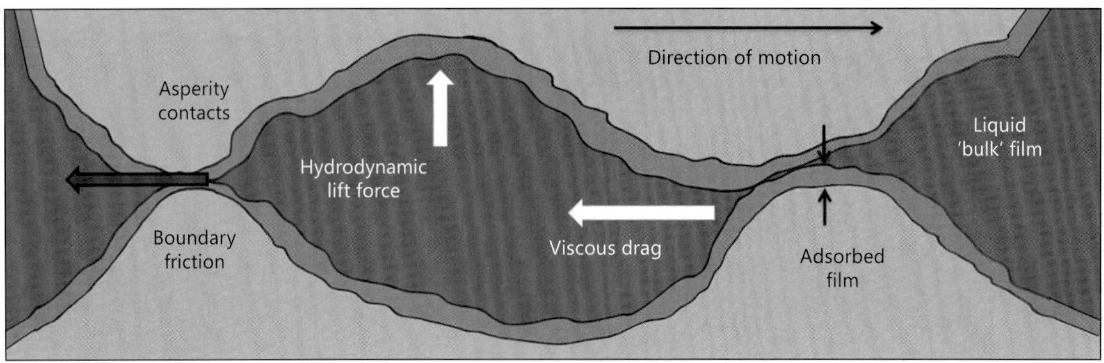

Fig. 1. Schematic diagram of a rubbing contact between rough surfaces in the presence of the lubricating polymer film and adsorbed polymer film.

preventing their motion. The magnitude of static friction can be prolific, as it has been illustrated in a stunt where a bridge formed by two paper phone books interleaved in every page was capable to withstand forces sufficient for lifting a small car off the ground [HeadSqeeze, MEGA Phonebook Car Lift – Power of Friction, http://www.youtube.com/watch?v = QNo8unbVtEo].

Clearly, the way to reduce friction is to keep surfaces apart, by which the momentum transfer and formation of adhesive contacts are minimised. Particularly effective at keeping surfaces apart are the repulsive electrostatic and magnetic forces. For example, the lubricity of solid lubricants such as graphite and molybdenum disulphide is attributable to a lamellar structure, comprising molecular sheets with very low cohesive forces between them. However, the applicability of these 'gap-maintaining' mechanisms in biolubrication is extremely limited, not the least due to the fact that water has an exceptionally high relative dielectric permittivity (the measure of attenuation of the electromagnetic field in the medium). In water, therefore, the range of these repulsive forces (e.g. negative van der Waals forces) is of the order of a few nanometres. Such a short range is absolutely insufficient to prevent asperity collisions in biological systems, and hence friction has to be minimised through alternative mechanisms.

Minimising friction in biological systems faces a set of challenges: Firstly in aqueous systems, the van der Waals forces are nearly always attractive, and hence at high enough loads the interaction between surfaces ultimately turns adhesive. Second, biological surfaces, such as the oral mucosa and the dental enamel, are rough. Therefore in the mouth the key mechanism for supporting lubrication is to keep a layer of liquid (saliva) between the surfaces to keep them at a safe distance (fig. 1).

The Concept of Length Scales in Oral Lubrication

Oral lubrication may range from oral processing of beverages, in which the liquid resides in the mouth only for a few seconds, to severe medical conditions associated with bruxism and some types of xerostomia, in which excessive abrasion of teeth or reduction in the saliva's ability to lubricate results in significant health impairment. Here we present a conceptual description of oral lubrication that addresses this diversity of real-life situations. The model of saliva's lubrication adopts accepted theories of different lubrication regimes from the areas of rheology, tribology and molecular polymer adsorption, and is based on a simple principle that the thickness of salivary film

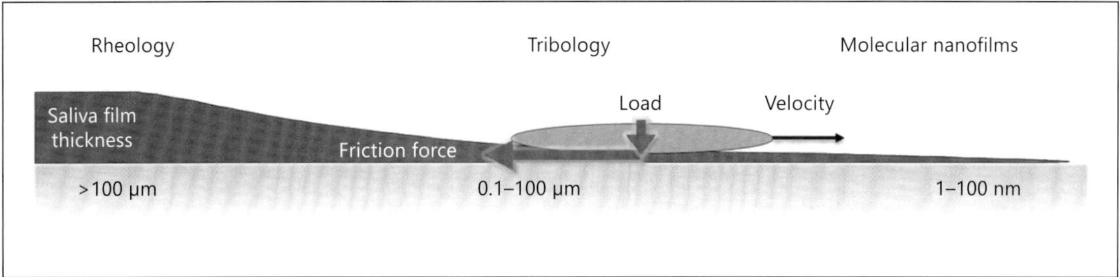

Fig. 2. A schematic representation of the changes in salivary film thickness during the rubbing motion between oral surfaces. In this cartoon, an oval represents the tongue sliding over the hard palate or a gum surface. At the onset of the motion, the salivary film thickness may be as large as a few 100 μm, yet during the progression of sliding this film thickness may decrease down to nanometre scale films. The typical domains of rheology, tribology and molecular nanofilm properties are noted above their respective thickness of a 'typical' lubricating film (adapted from Stokes et al. [3]).

is not constant during the rubbing process in the mouth. Instead, it changes spanning multiple length scales from a few hundred micrometres to as thin as a few tens of nanometres (fig. 2).

Full-Film Lubrication

If surfaces move at high enough speeds, like in engines or ball bearings, it is advantageous to have a viscous liquid such as oil or grease being placed in a rubbing contact. The viscous drag exerted by the viscous fluid creates a lift force that keeps surfaces apart, minimising dynamic friction. This effect is called hydrodynamic lubrication (fig. 3).

The effect of the hydrodynamic lift is the principle behind waterskis, hydrofoils and airplane wings. In the simplest case the lift force is proportional to the fluid viscosity, fluid speed in the gap and the geometry of a rubbing contact. In such systems the friction force is determined by the balance between two forces:
- the lift force necessary to separate surfaces and avoid asperity contacts;
- the drag force produced by the sheared fluid.

As speed increases, the drag force increases faster than the lift force, and therefore the friction force increases proportionally with the fluid ve-

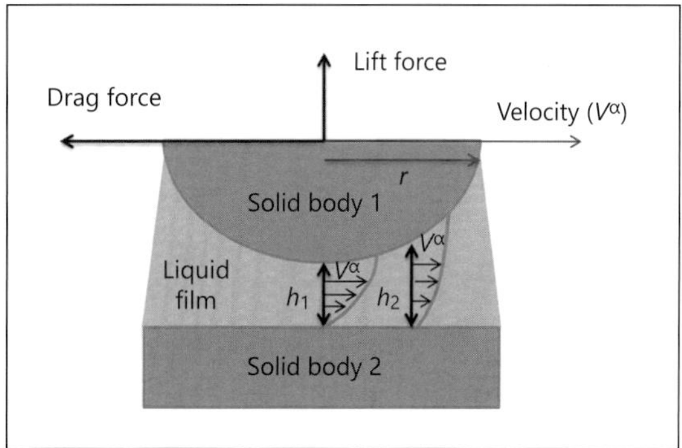

Fig. 3. A lift force effect induced by a viscous drag around a spherical probe. Since the gap formed by curved surfaces changes with radial position h = h(r), the hydrodynamic pressure distribution is not constant, but changes from the highest values in the areas of minimal gap to the lower values on the periphery. This pressure gradient is directly proportional to the lift force.

locity. By the same token, the friction force is at its minimum when the thickness is just about to surpass the size of asperities.

If the contacts are formed by elastic solids, the fluid pressure can deform surfaces, thereby modifying the height profile within the gap. Since the thickness of the liquid film becomes dependent on deformation, it shifts the balance of the hydrodynamic drag and the lift force thereby delaying

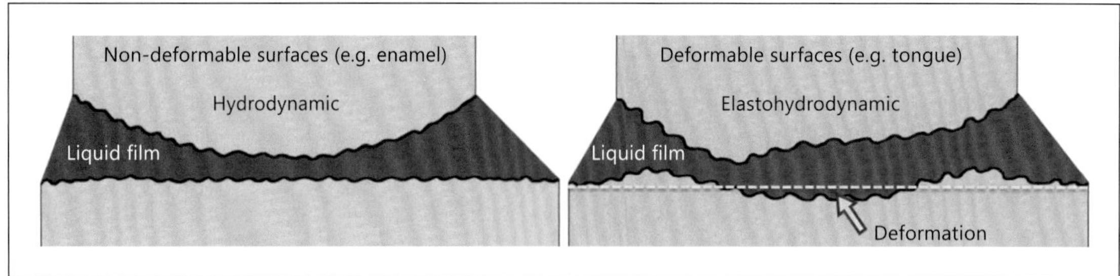

Fig. 4. The comparison of the gap between rubbing surfaces formed between hard/non-deformable (hydrodynamic lubrication) and deformable surfaces (elastohydrodynamic lubrication). In the oral lubrication, the closest analogues would be the rubbing contacts formed between teeth and soft oral surfaces, respectively.

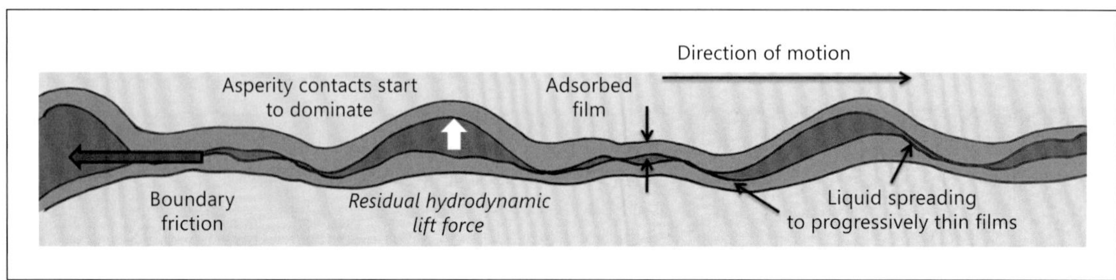

Fig. 5. Schematic illustration of mixed lubrication regime, where residual liquid film coexists with asperity contacts. Under these conditions the liquid film can be squeezed down to very thin layers (of the order of a few hundreds of nanometres) properties of which strongly depend on the liquid's spreading behaviour.

the onset of mixed lubrication. This has a particularly significant effect when the gap is only slightly larger than the size of asperities. The lubrication under such conditions is termed elastohydrodynamic lubrication (fig. 4). Since most of the oral surfaces except the teeth are soft, the elastohydrodynamic regime has the dominant effect in the full-film saliva lubrication.

Mixed Lubrication

When speeds are low or loads are high, the hydrodynamic lift force can no longer keep surfaces apart, and therefore the asperity contacts will start to form, while the lubricating liquid gets squeezed out from the gap. The term 'mixed' re-flects the fact that the total friction force is a result of contributions from the elastohydrodynamic part (coming from the residual liquid in the gap) and the asperity contact friction (fig. 5). Since the friction force between asperities is usually higher than the elastohydrodynamic component, the net friction force increases with decreasing effective film thickness. At this stage the surface wetting behaviour of the fluid becomes an important factor that promotes fluid entrainment into the gap.

Boundary Friction

As speed decreases further, the total area of asperity contacts increases proportionally to a point when viscous drag fails to produce the required

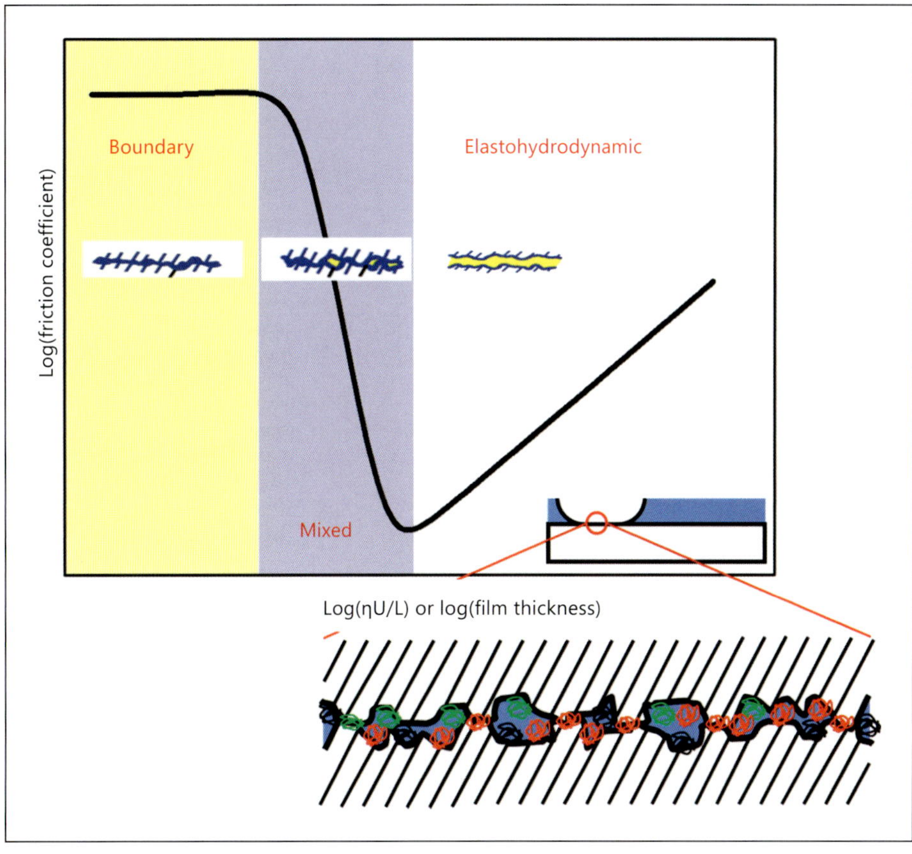

Fig. 6. Stribeck curve, a function of friction coefficient plotted against the parameter ηU/L or film thickness (in elastohydrodynamic regime) plotted in logarithmic coordinated, where η is viscosity, U is entrainment speed and L is load (adapted from de Vicente et al. [5]).

lift force and the area of asperity contacts becomes directly proportional to the applied force. As a result, the friction force becomes independent of sliding speed. At this stage, called the boundary friction, the friction force is totally dominated by the friction between asperity contacts. In biological systems, water is a key lubricant, because it stays liquid even down to molecularly thin films [4], which is opposite to many other liquids that tend to get more structured when confined. However, without any additional structural support, water can be readily squeezed out from the gap allowing bare surfaces to engage into the direct contact. The role of adsorbed films of proteins and glycoproteins is in providing such structural support and thus retaining water in the gap.

Stribeck Curve

This multistage dependency of the friction force on the speed can be formalised using the Stribeck description, which is universally applicable for rubbing contacts in the presence of liquid lubricants. Figure 6 presents a schematic Stribeck curve that is the dependency of the friction coefficient

$$\mu = \frac{friction\ force}{load}$$

on the Stribeck parameter

$$S = \frac{fluid\ velocity \cdot fluid\ viscosity}{load}.$$

In the full-film regime, the Stribeck parameter is directly proportional to the gap between the surfaces. In the mixed regime this dependency is more complex, and it is possible to relate S to an 'effective gap' or the density of the asperity contacts in the gap. In the boundary regime, the gap stays constant irrespective of S, with any deviations being indicative of changes in surface layer microstructure or presence of particulates such as food particles or microscopic pieces of calculus.

In the oral lubrication, a single stroke of the tongue can drive saliva from a thicker film (elastohydrodynamic lubrication) to a thinner film (boundary friction). This process is a continuum of different lubrication regimes that follows a path along the Stribeck curve. By swirling a spoonful of vegetable oil around the mouth, one can experience a relatively clear-cut elastohydrodynamic lubrication, which will dominate prior to the swallowing. On the other hand, teeth grinding is a vivid manifestation of boundary friction, with the characteristic sound being a sign of energy dissipation. The Stribeck description enables to define the process of oral lubrication using a single framework that encompasses all aspects of saliva physical and physicochemical properties, such as rheology, interfacial tension, surface forces and protein adsorption.

Full Film Lubrication and Saliva Rheology

At the initial stages of salivary lubrication, when film is thick (>1 μm), the lubrication mechanism relies on the hydrodynamic lift force. In the classical lubrication theory, the primary parameter of

concern was viscosity [6], the resistance against flow. Viscosity (η) is an intrinsic material property defined as the ratio of the force per unit area (or shear stress σ) required to shear liquid at the rate of a unit ($\dot{\gamma}$):

$$\eta = \frac{F/A}{\dot{\gamma}} = \frac{\sigma}{\dot{\gamma}}.$$

On a molecular level, viscosity is related to the friction force between electron clouds of atoms and molecules, and for many simple liquids the process of molecular collisions happens so fast that no matter how fast the bulk liquids shear the force required stays linear with the rate. Hence for such so-called newtonian liquids viscosity is independent of shear rate.

Polymer solutions (including saliva) may often display non-newtonian behaviour such as for instance shear thinning, i.e. their viscosity decreases with increasing shear rate. This occurs due to alignment of polymer chains along the direction of the flow that reduces hydrodynamic resistance and results in shear stress being constant regardless of the shear rate. Long polymers not only align along the direction of the flow, but can also be stretched by the shear. Like a spring being stretched, this produces a force that pushes the liquid back. This 'push-back' of the polymer chains together with the liquid bound within solvated shells results in the so-called elastic effect, which provides resistance against deformation. Since viscous flow and elastic effects occur simultaneously under shear, this rheological behaviour is defined as visco-elasticity. The 'push-back' process of the extended polymer chains is characterised by the time parameter – a relaxation time (λ), i.e. how quickly the structure relaxes back. If the structure of the polymer is complex (e.g. a complex coil) or the polymer solution is polydisperse, a number of relaxation time parameters or their distribution is necessary to describe the visco-elastic behaviour of such systems.

For some types of polymers this elastic response acts not only in the direction of the shear force, but spans into the 'third' dimension and acts perpendicularly to the direction of flow, thus pushing surfaces apart. Because this effect is secondary to the shear-induced polymer extension, it comes as a display of non-linear visco-elastic behaviour, and conventionally in the rheological literature is termed 'primary normal stress difference' (N_1) that acts perpendicularly to the direction of the shear. This normal force creates an additional load-bearing component to the hydrodynamic lift that pushes surfaces apart and therefore is highly advantageous for facilitating elastohydrodynamic lubrication [7]. This additional force may be actually greater than the viscous drag force. Such liquids become very effective in organising the lift force necessary for keeping surfaces apart at relatively low shear viscosity. This will achieve thinner films and hence even lower values of friction forces (fig. 6).

Saliva rheological behaviour is characterised by both shear thinning and a very high elastic component [7]. Indeed, saliva is a highly non-newtonian fluid with large relaxation times probably due to the presence of extremely large polymeric aggregates that relax very slowly. This slow relaxing elasticity manifests itself macroscopically during, for example, extensional deformation when very well-known saliva strings are forming (fig. 7).

The ratio of normal stress and shear stress (N_1/σ) is a useful parameter for describing non-newtonian behaviour of saliva, as it captures the balance between the work required to stretch the polymer molecules and the elastic energy they store. Saliva that is collected by stimulation with acid has a stress ratio of 100, a value that is unmatched by most of the polymer fluids, which typically stay below 10. This very high value appears even more striking, when taking into account the very low viscosity of saliva. This unique combination of rheological properties, low viscosity, shear thinning and high elasticity, is most likely responsible for the exceptional elastohydrodynamic (full-film) lubricating properties of saliva. Due to its high stress ratio [7], a relatively

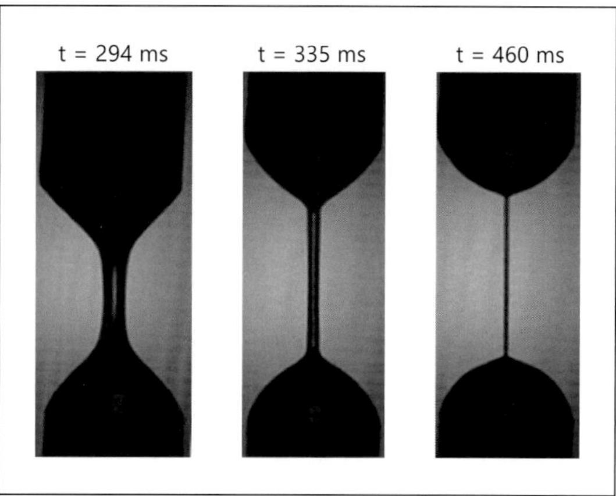

Fig. 7. Extensional deformation of unstimulated (resting) saliva as a function of time. The photographs were taken during self-thinning of the saliva liquid thread (adapted from Punyadeera et al. [66]).

thin salivary film is able to keep opposing surfaces apart. This effect impedes the formation of asperity contacts between rubbing surfaces, and hence defers the onset of the mixed lubrication. At the same time, the low viscosity combined with shear thinning reduces the hydrodynamic drag force resulting in low friction.

The magnitude of the elastic component strongly depends on the type of saliva and the way it has been collected. It is extremely susceptible to saliva aging and any treatments, such as centrifugation, homogenization dilution or filtering [8]. Interestingly, acid-stimulated saliva has an approximately 10 times higher stress ratio compared to mechanically stimulated saliva, which in turn is much higher than what would be expected for a polymer solution with a viscosity close to water. This dependency on stimulation is probably due to the difference in elastic properties of secretions originating from different glands, with secretions from submandibular and sublingual glands having a stronger contribution to elasticity. The very same secretions have also the higher mucin content. In terms of rheological behaviour, the salivary mucinous components can roughly be split into three fractions, gel MUC5B, sol MUC5B and MUC7 [9]. The likely candidate being responsible for the observed rheological behaviour is the gel MUC5B fraction that comprises supramolecular aggregates of large-molecular-weight mucin molecules joined in a filamentous network that may also be qualified as a weak gel [10]. These are exactly the type of molecular structures that display non-linear elastic behaviour as experimentally observed for saliva. The formation of supramolecular aggregates is a key feature of mucins and involves a number of interactions such as disulphide bridging, hydrophobic and electrostatic colloidal interactions. A delicate equilibrium exists between associated and weakly associated mucins, which exist as 'dangling chains' within the matrix [10, 11]. This complex pattern of interactions as well as the presence of soluble fractions such as MUC7 is responsible for a broad spectrum of relaxation time that characterises polymer relaxation. Another important rheological property of saliva is its high resistance against elongation (so-called extensional viscosity, illustrated in fig. 4). The ratio of extensional viscosity to shear viscosity is expressed as Trouton's ratio. Newtonian liquids have a Trouton's ratio of 3. Highly elastic polymer solutions can have Trouton's ratios as high as $\geq 1,000$. For centrifuged saliva a Trouton's ratio of up to 120 has been measured, which is a marked display of its non-newtonian behaviour. However, the reported values might be an underestimation, taking into account that the sample preparation methods are likely to result in partial depletion of mucins. The break-up of a salivary filament often undergoes the formation of beads-on-a-string structures [12] that were observed also with other mucous systems [13], suggesting the transience of saliva elastic properties. It suggests that after extension the polymeric species may either break up or collapse resulting in the reduction of both viscosity and relaxation time. Indeed, saliva is known to lose much of its elastic behaviour shortly after expectoration, thus suggesting that the mucin microstructure is highly fragile and may exist very shortly after secretion of mucins. Already during the passage in the ducts, the mixing with the serous secretion occurs. Upon excretion into the oral cavity, the mucin-rich submandibular/sublingual secretions get diluted even further by the watery parotid secretions. During the course of aging of the sample, one can observe changes in the microstructure including the formation of proteinaceous aggregates with mixed composition. The latter, unlike fibrous networks, contribute very little to the elasticity of the fluid.

To date there is no unequivocal proof of what components of saliva give rise to its elastic behaviour. Mucins are prominent polymeric species and therefore are likely to be key to the saliva's elastic behaviour; however, the exact type of mucins and the role of the assemblies with lower-molecular-weight proteins still remain to be elucidated. A number of studies showed that purified MUC5B mucins do not replicate the rheological properties of saliva. The addition of Ca^{2+} may facilitate the gelation, but not necessarily lead to the increased elastic response. Another factor is the role of hydrophobic interactions that in the absence of low-molecular-weight proteins may result in extensive self-association of mucin molecules leading to the loss of a filamentous and extended conformation [14].

Mixed Lubrication and Saliva Interfacial Elasticity

The tongue surface is covered by papillae with submillimetre roughness. Such a high roughness renders fully developed elastohydrodynamic lubrication unattainable (unless one swirls viscous oil). By the same virtue, saliva is always trapped within the tortuous structure of the tongue and hence cannot be completely squeezed out to give way to boundary friction (unless saliva production is impaired). This leaves rubbing contact between the tongue and other oral surfaces to be mostly in the mixed lubrication regime.

In the mixed regime, the liquid films become thinner, and the spreading of the film takes up the leading role in facilitating lubrication. The spreading process is governed by the surface tension. However, the surface tension, defined as energy required to form a unit area of interface, is often insufficient to describe dynamic spreading processes. Firstly, since the surface tension is related to the adsorption of the surface active molecules at the interface, it is constrained by the time required for the adsorption to take place. Another factor is related to the polymeric nature of the networks assembled at the interface. Besides the energy required moving a molecule from the bulk to the surface, an additional energy component is associated with the stretching of the polymer network. By analogy with bulk rheology, this energy can be split into two components: elasticity and viscosity. Interfacial elasticity is particularly high for crosslinked or crystallised proteins such as hydrophobins [15, 16]. Interfacial rheological properties play important roles in spreading behaviour, as they are responsible for the deformation of the interface when the liquid navigates itself through the tortuous network of capillaries formed by the asperity contacts.

The interfacial rheological properties of saliva, for instance at the air-saliva interface, are no less unique than its bulk rheological properties. The phenomenon responsible for such uniqueness is linked to the formation of an elastic molecular film at the air-saliva interface. Proctor et al. [17] and Hamdan et al. [18] demonstrated that in parotid saliva the major component of this proteinaceous film is statherin, a surfactant-like protein. However, purified solutions of statherin do not replicate elastic properties of saliva at the air-saliva interface, suggesting contribution from other proteins. Kazakov et al. [19] have found that the interfacial elastic modulus of the sublingual and submandibular secretions is actually higher than that of the parotid secretions, with whole-mouth saliva having inter-mediate values. This suggests that mucins may play a role in the formation of the elastic layer at the air-saliva interface. In such a scenario, surfactant-like statherin would form an interfacial layer, whilst mucin would reinforce it by physicochemical interaction, thereby giving rise to elasticity.

Boundary Friction and Adsorbed Salivary Pellicle

When the liquid film is largely squeezed out from the contact zone, the layer of proteins (salivary pellicle) bound to the oral substrates takes on the role of supporting the applied load and modulating the boundary friction. The architecture of the adsorbed salivary pellicle due to its hydration and load-bearing capacity is key to lubrication [20]. The load-bearing capacity stems from the steric repulsion between adsorbed salivary layers [21–23] (fig. 8).

These layers display a degree of plasticity with decompression curves showing a shorter range of forces than compression ones. This phenomenon, called hysteresis, indicates a slow relaxation of surface structure upon mechanical deformation, and hence hints to a possible water movement in and out of the film. The adhesive forces between salivary films are dominated by bridging adhesion (fig. 9) as it was evident from the dependency of adhesion on the time surfaces are held in contact [24]. Such bridging has also been found in mucin pellicles [25, 26], which suggests that load-bearing capacity is highly dependent on the cohesiveness within the salivary pellicle, and that lubrication can be severely impaired if adhesive interactions are not suppressed.

The Effect of Roughness of Boundary Friction

The friction coefficients (μ) for saliva and saliva mimics range between 0.01 and 0.45, depending on the conditions of the experiment. The lowest

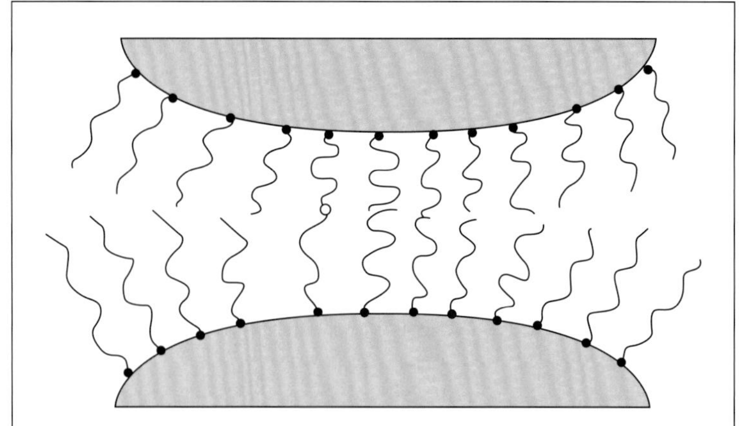

Fig. 8. Effect of steric repulsion on interaction between surfaces. The presence of polymer brushes on the surface provides an additional repulsive barrier that contributes to the load-bearing capacity and helps reduce friction.

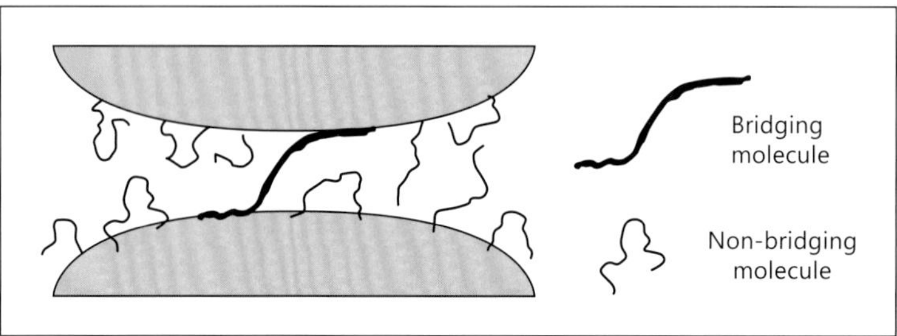

Fig. 9. Effect of bridging adhesion. When surfaces are close to each other, the adsorbed molecules may form bridges between surfaces that contribute to the adhesion and consequently higher friction force (adapted from Harvey et al. [26]).

estimate is therefore at least 2 orders of magnitude lower than the friction measured on dry or water-lubricated hydrophobic contacts. The friction coefficient of a smooth saliva-lubricated polydimethylsiloxane contact was found to be $\mu = 0.02$ [27], with somewhat similar results being found for the salivary MG1 (MUC5B) and MG2 (MUC7) fractions [28, 29]. Berg et al. [22] performed atomic force microscopy colloidal probe experiments using a smooth silica colloidal sphere sliding on a silica substrate, lubricated by a 10% human whole-saliva solution. They obtained a friction coefficient of $\mu = 0.03$, which is very similar to the above values. However, when biological surfaces are used, e.g. the tongue surface, the values of the friction coefficient are usually higher. Ranc et al. [30] reported $\mu = 0.16$ in the friction measurements performed on the tongue at 55–65% relative humidity. Prinz et al. [31] used a custom-made rig that uses pig's tongue and oesophagus as one of the rubbing surfaces. In this work the reported values of μ were in the range between 0.1 and 0.35, depending on load and speed. It appears that roughness is a crucial factor in friction measurements, and these conditions should be carefully distinguished. The effect of roughness was clearly shown using polydimethylsiloxane compliant surfaces; on rough surfaces (root mean square roughness approx. 380 nm), the friction coefficient of the order of 0.1 was measured, while on the smoother surfaces (root mean square roughness approx. 8 nm) a friction coefficient of 0.02 [27] was measured.

Boundary Friction of Molecularly Thin Pellicles

The key boundary lubricant in the oral cavity consists of water, hydrated ions and ion groups of proteins that must be strongly attached to the adsorbed layer yet retain mobility in order to support lubrication [4]. The mucin and protein network effectively traps water and ions within the pellicle, thus ensuring the first requirement is met, while open architecture effects the enhanced mobility. The surface chemistry has a significant influence on saliva lubrication, especially when mixed lubrication and boundary friction are considered. For hydrophilic surfaces, which are generally more lubricious due to low adhesion, salivary film does little to facilitate boundary lubrication [29, 32, 33]. This is also because salivary pellicles formed on hydrophilic substrates are much thinner compared to hydrophobic substrates. Another aspect of lubrication is related to the soft nature of many oral surfaces. The pressure within rubbing contacts formed by soft tissues is rarely in excess of a few tens of atmospheres [34], which exerts little abrasion and which does not affect the integrity of the salivary film [35]. The addition of surface active ingredients can result in dislodgement of the lubricating films from the surface, and consequently loss of lubrication [27, 36]. Despite a number of factors that can compromise the effectiveness of the lubricious layer, it displays remarkable robustness as illustrated by the constant low friction values that have been obtained over time in a continuous rubbing experiment [27]. It was also possible to dry and then rehydrate the lubricious film without permanent loss of lubrication.

A hunt for a molecule behind saliva lubrication was nothing less than exhaustive. At different times, mucins, proline-rich proteins and statherin have been suggested as having a leading role in salivary lubrication. Lee et al. [37], for example, obtained boundary friction coefficients of $\mu =$ 0.02 for pig gastric mucins at a pH = 2. Under neutral pH conditions, a friction coefficient for pig gastric mucins of approximately 0.1 was measured [38, 39]. The result of Lee et al. is most likely an effect of gelation of pig gastric mucins under low pH conditions, which effectively results in a gel-gel friction, which may not be relevant for the oral environment. An early work by Douglas et al. [40] examined lubrication between the polished enamel surfaces mediated by saliva and salivary fractions. It was found that statherin purified from saliva yielded the lowest values of friction coefficient of approximately 0.4, as compared to the whole-saliva or mucin-rich fraction which was ≥ 0.7. However, this work had poor definition of the surface roughness, which renders accurate evaluation of the applied pressure impractical. Berg et al. [41] and Harvey et al. [42] found that purified statherin has actually a very marginal capacity to promote lubrication on non-enamel substrates with friction coefficient values of $\mu \sim$ 0.66 and $\mu \sim$ 0.15, respectively. At the same time the authors found that purified acidic proline-rich proteins displayed the lowest friction coefficient even superseding that of whole saliva ($\mu <$ 0.03). However, results obtained on filtered saliva samples [21] that contain physiologically relevant quantities of acidic proline-rich proteins but are depleted of mucins, demonstrated that in the absence of mucins saliva has a very limited capacity to lubricate ($\mu \sim$ 0.24). To date there is no full consensus regarding detailed mechanisms underpinning a boundary regime in oral lubrication. However, it is likely to be a combination of mucins/glycosylated proline-rich proteins and lower molecular weight proteins forming a hydrated visco-elastic composite layer [43–45].

The investigation of individual proteins shed some light on salivary lubrication; with the main message being that no single component is sufficient to explain the phenomenon. Lundin et al. [45] showed that the combination of mucin with bovine serum albumin yielded a lower friction than any of the components taken individually, suggesting the importance of synergistic effects.

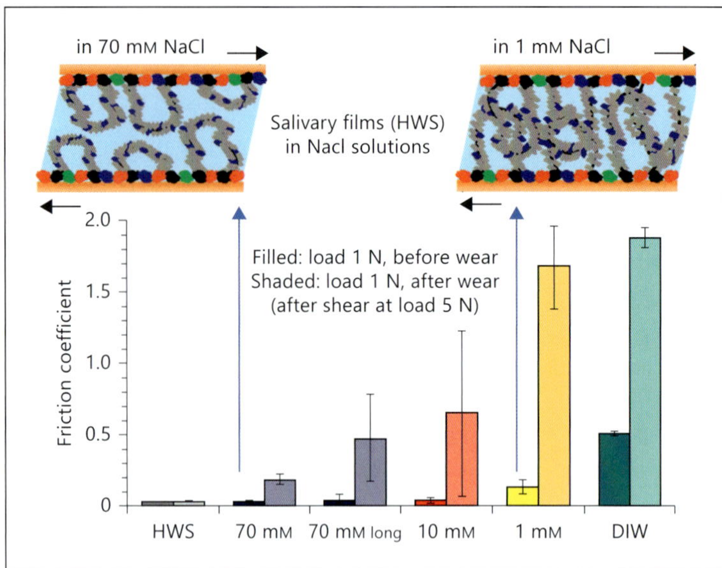

Fig. 10. Dependency of saliva friction coefficient on ionic strength and history of high load abuse (adopted from Macakova et al. [43]). HWS = Human whole saliva; DIW = deionised water.

A recent investigation of saliva lubrication using a surface force apparatus [21] and an atomic force microscope [46] demonstrated that the nanoscale lubrication by saliva adsorbed on hydrophilic substrates, mica and tooth enamel, respectively, is modest compared to the reduction observed on hydrophobic substrates.

The investigation of salivary film lubrication between hydrophobic substrates showed a strong dependence on ionic environment. The ionic strength has a profound impact on the architecture of the salivary pellicle, as it was demonstrated by Macakova et al. [47]. The main impact of ionic strength is on the hydration and visco-elasticity of a salivary film, while the adsorbed mass of proteins remains largely unchanged. Decreasing salt concentration from physiological 70 to 1 mM causes the pre-adsorbed film to reversibly swell. However, exposure of the salivary film to deionised water causes the film to irreversibly collapse due to the onset of attractive electrostatic interactions between glycosylated groups of the extended chains of the upper layer and proteins that form the inner layer. These changes in the structure had a dramatic impact on the lubrication properties, and particularly on the wear resistance [43].

Figure 10 shows the friction coefficients measured between salivary films under different ionic strength conditions before and after the film had been abused at higher loads. Whole saliva displays a great deal of resilience with the friction coefficient remaining unchanged after the film has been subjected to the wearing conditions under elevated loads. The exchange of saliva to a solution with the matching ionic strength (70 mM NaCl) results in no loss of lubrication observed under low loads, but the impact of wear becomes apparent. The longer the rinsing time, the more such vulnerability was exposed. This suggests that partial exchange with the bulk film is likely to be essential for the film repair mechanism. When salivary film was exposed to the reduced ionic strength, some loss of layer elasticity as well as its extension was observed in a quartz crystal microbalance with dissipation monitoring experiment. These changes resulted in a nearly complete loss of wear resistance, as evidently it was for the case of 1 mM NaCl buffer.

In addition to changes in the ionic strength, the integrity of the adsorbed salivary film can be

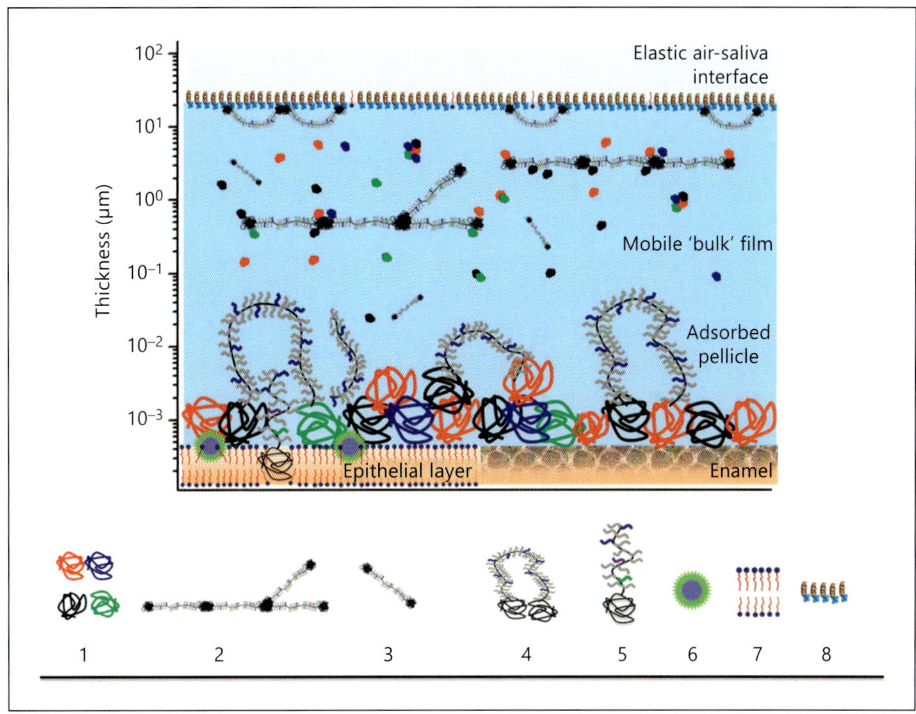

Fig. 11. Schematic 3-dimensional model of a salivary film formed on soft and enamel surfaces. 1 = Low-molecular-weight salivary proteins, proline-rich proteins, cystatins, statherin, secretory IgA etc.; 2 = MUC5B mucin supramolecular aggregates; 3 = MUC7 mucins; 4 = surface-bound MUC5B/MUC7 mucins; 5 = membrane-bound mucins; 6 = transglutaminase crosslinks; 7 = phospholipid lipid bilayer; 8 = interfacial statherin film. The coloured oligosaccharide chains of mucins represent charged (sialic acid terminated and sulphated residues) and N-glycan fragments.

compromised by a number of factors, including surfactants, dietary polyphenols or mechanical action [35, 36, 48]. For example, the elution of salivary pellicle with sodium dodecyl sulphate results in the loss of salivary lubrication [27], which is due to the removal of the hydrated part of salivary adsorbed film, as it was shown in the adsorption studies by Santos et al. [49].

Saliva Lubrication and the Molecular Structure of Salivary Films

The saliva pellicle, a bound proteinaceous layer, is formed on all tissues in the oral cavity [50]. It plays a key role in the maintenance of oral health, and apart from lubrication it provides hydration and immune response, shapes the microbial flora population and regulates tooth mineralization processes [51, 52]. The composition and structure of salivary pellicle depend on location in the oral cavity and the nature of the underlying oral substrates, i.e. mineralised, keratinised and non-keratinised tissues, as well as hydrophobic versus hydrophilic areas [53]. Despite the diversity, there are key features that are captured in a schematic diagram representing a 3-dimensional model of a salivary film (fig. 11). The film has roughly 5 layers: the precursor layer, the tightly bound protein layer, the layer of the surface-anchored mucin brush, the liquid bulk layer and air-saliva interface layer.

Salivary proteins have the ability to bind directly to the enamel or mucosa epithelial layer [54]. This process is often triggered by precursor pellicle proteins. This is seen within the enamel pellicle, where proteins such as statherin and proline-rich proteins are thought to initiate pellicle formation [55]. On soft tissues, the evidence suggests that mucin-mucin interaction between salivary MUC5B [56] and membrane-bound MUC1 is a likely trigger mechanism of the initial stages of mucosal pellicle formation [57]. There is evidence that covalent links, mediated by transglutaminase, which catalyse crosslinks between glutamine and lysine residues, may contribute to the attachment of this initial precursor film to the epithelium; however, this may not be a universal mechanism [58].

Further the gradual build-up of the adsorbing salivary proteins results in the formation of the tightly bound protein layer that has a multilayered structure due to a complex pattern of intramolecular interactions. The composition of this layer of the pellicle reflects to some extent that of the whole-mouth saliva, with some proteins being either over- or underrepresented [59, 60; see the paper by Lindh and Arnebrandt, this vol., pp. 30–39]. For example, in the enamel pellicle proline-rich proteins are typically overrepresented, while amylase is relatively underrepresented [61]. The dominant constituents of this layer are proline-rich proteins (primarily acidic), cystatins, histatins and statherin. The latter has a particular affinity to the hydroxyapatite of tooth enamel due to calcium-binding domains. Other synergistic interactions such as heterotypic complexes formed between secretory IgA and lactoferrin with MUC7 contribute to improved protein binding within the pellicle, and immunoprotective properties of the mucosal pellicle layer [62].

The properties of these precursor and tightly bound layers are key to the attachment of the pellicle to the oral substrate and hence impact its wear resistance.

During the protein assembly process, alongside with lower-molecular-weight proteins, the non-glycosylated parts of MUC5B are getting trapped and anchored within this layer. With little room to manoeuvre, the slowly diffusing glycosylated parts of mucins become protruded into the liquid salivary film, thus resulting in the formation of a highly hydrated and diffuse outer layer. The structure of this layer bears similarity with surface brushes and may also contain loops of mucin chains with liquid water trapped within. Such an architecture is key to lubrication as it results in a highly hydrated and thick (up to 100 nm) layer with high load-bearing capacity due to an increased visco-elastic response of the layer [43, 47]. Thus, it is an effect of co-adsorption of low- and high-molecular-weight proteins that results in a multilayer assembly that provides both strong adhesion to the oral surfaces and hydration and lubrication functionality. This structural model has been corroborated in a number of studies using saliva as well as model systems [44, 45]. The constant exchange of material with the bulk of the salivary films keeps the pellicle regenerated and maintains its integrity [63].

Beyond the adsorbed pellicle, salivary film has a bulk salivary layer that is characterised by the presence of mucin aggregates that give rise to saliva's non-newtonian rheological properties. Finally, an elastic proteinaceous layer at the interface with the air completes the structure of the salivary film. This multilayered structure contributes to the entire spectrum of saliva's lubrication functionality and makes it a versatile lubricant capable of supporting different lubrication regimes in a broad spectrum of length and time scales.

Future Perspective

The salivary film is capable of withstanding a broad range of stresses, yet due to its simplicity and self-assembly it can be sacrificed if needed and thereafter quickly replenished. The structural properties of salivary film are a product of a finely tuned balance between intramolecular interactions and kinetic factors. This balance, however, can be shifted. The properties of the film may be sensitive to the environment and can reflect several oral pathological states such as erosion and periodontal disease [60]. Such factors as diet, circadian cycles, microbial flora, state of hydration, medication and psychophysiological conditions (e.g. stress) influence the formation and properties of the salivary pellicle [64] and ultimately impact its ability to lubricate. For example, patients with some types of dry mouth conditions may see an increase in the protein content of the pellicle [65]. This change may shift the assembly of the salivary film towards more densely packed proteinaceous film that bears a weaker capacity to support load and lubricate compared to an open and hydrated architecture of the healthy pellicle. In gingivitis, one can observe the release of the excessive amount of albumin that likewise can shift the balance of protein assembly rendering saliva lubrication.

Further research effort is required to understand the details of salivary film assembly. The development of this area is key to creating healthier foods and advanced saliva substitute systems. One can also envisage the development of novel medical treatments based on the template protein self-assembly that can help alleviate the symptoms associated with the loss of lubrication.

Acknowledgement

I would like to formally acknowledge the editorial board (E.C.I. Veerman and A.J.M. Ligtenberg) for their contribution and invaluable guidance in the publication of this paper. I would also like to thank Prof. Jason Stokes and Dr. Lubica Macakova for helpful discussions.

References

1 Berkovitz BKB: Oral Biology. Edinburgh, Churchill Livingstone, 2011.

2 Breslin PAS, Gilmore MM, Beauchamp GK, Green BG: Psychophysical evidence that oral astringency is a tactile sensation. Chem Senses 1993;18:405–417.

3 Stokes JR, Boehm MW, Baier SK: Oral processing, texture and mouthfeel: from rheology to tribology and beyond. Curr Opin Colloid Interface Sci 2013;18:349–359.

4 Gaisinskaya A, Ma L, Silbert G, Sorkin R, Tairy O, Goldberg R, et al: Hydration lubrication: exploring a new paradigm. Faraday Discussions 2012;156:217–233.

5 De Vicente J, Stokes JR, Spikes HA: Soft lubrication of model hydrocolloids. Food Hydrocolloids 2006;20:483–491.

6 Sajewicz E: Effect of saliva viscosity on tribological behaviour of tooth enamel. Tribol Int 2009;42:327–332.

7 Stokes JR, Davies GA: Viscoelasticity of human whole saliva collected after acid and mechanical stimulation. Biorheology 2007;44:141–160.

8 Schulz BL, Cooper-White J, Punyadeera CK: Saliva proteome research: current status and future outlook. Crit Rev Biotechnol 2013;33:246–259.

9 Wickstrom C, Hamilton IR, Svensater G: Differential metabolic activity by dental plaque bacteria in association with two preparations of MUC5B mucins in solution and in biofilms. Microbiol SGM 2009;155:53–60.

10 Taylor C, Allen A, Dettmar PW, Pearson JP: The gel matrix of gastric mucus is maintained by a complex interplay of transient and nontransient associations. Biomacromolecules 2003;4:922–927.

11 Taylor C, Draget KI, Pearson JP, Smidsrod O: Mucous systems show a novel mechanical response to applied deformation. Biomacromolecules 2005;6:1524–1530.

12 Bhat PP, Appathurai S, Harris MT, Pasquali M, McKinley GH, Basaran OA: Formation of beads-on-a-string structures during break-up of viscoelastic filaments. Nat Phys 2010;6:625–631.

13 Celli J, Gregor B, Turner B, Afdhal NH, Bansil R, Erramilli S: Viscoelastic properties and dynamics of porcine gastric mucin. Biomacromolecules 2005;6:1329–1333.

14 Bromberg LE, Barr DP: Self-association of mucin. Biomacromolecules 2000;1:325–334.

15 Cox AR, Cagnol F, Russell AB, Izzard MJ: Surface properties of class II hydrophobins from Trichoderma reesei and influence on bubble stability. Langmuir 2007;23:7995–8002.

16 Stanimirova RD, Gurkov TD, Kralchevsky PA, Balashev KT, Stoyanov SD, Pelan EG: Surface pressure and elasticity of hydrophobin HFBII layers on the air-water interface: rheology versus structure detected by AFM imaging. Langmuir 2013;29:6053–6067.

17 Proctor GB, Hamdan S, Carpenter GH, Wilde P: A statherin and calcium enriched layer at the air interface of human parotid saliva. Biochem J 2005;389:111–116.

18 Hamdan S, Proctor GB, Carpenter GH: Surface active protein in human parotid saliva. J Dent Res 2001;80:1168.

19 Kazakov VN, Udod AA, Zinkovych II, Fainerman VB, Miller R: Dynamic surface tension of saliva: general relationships and application in medical diagnostics. Colloids Surfaces B Biointerfaces 2009;74:457–461.

20 Dedinaite A: Biomimetic lubrication. Soft Matter 2012;8:273–284.

21 Harvey NM, Yakubov GE, Stokes JR, Klein J: Lubrication and load-bearing properties of human salivary pellicles adsorbed ex vivo on molecularly smooth substrata. Biofouling 2012;28:843–856.

22 Berg ICH, Rutland MW, Arnebrant T: Lubricating properties of the initial salivary pellicle – an AFM Study. Biofouling 2003;19:365–369.

23 Hannig M, Dobbert A, Stigler R, Muller U, Prokhorova SA: Initial salivary pellicle formation on solid substrates studied by AFM. J Nanosci Nanotechnol 2004;4:532–538.

24 Schwender N, Huber K, Al Marrawi F, Hannig M, Ziegler C: Initial bioadhesion on surfaces in the oral cavity investigated by scanning force microscopy. Appl Surface Sci 2005;252:117–122.

25 Efremova NV, Huang Y, Peppas NA, Leckband DE: Direct measurement of interactions between tethered poly(ethylene glycol) chains and adsorbed mucin layers. Langmuir 2002;18:836–845.

26 Harvey NM, Yakubov GE, Stokes JR, Klein J: Normal and shear forces between surfaces bearing porcine gastric mucin, a high-molecular-weight glycoprotein. Biomacromolecules 2011;12:1041–1050.

27 Bongaerts JHH, Rossetti D, Stokes JR: The lubricating properties of human whole saliva. Tribol Lett 2007;27:277–287.

28 Aguirre A, Mendoza B, Reddy MS, Scannapieco FA, Levine MJ, Hatton MN: Lubrication of selected salivary molecules and artificial salivas. Dysphagia 1989;4:95–100.

29 Aguirre A, Mendoza B, Levine MJ, Hatton MN, Douglas WH: In vitro characterization of human salivary lubrication. Arch Oral Biol 1989;34:675–677.

30 Ranc H, Elkhyat A, Servais C, Mac-Mary S, Launay B, Humbert P: Friction coefficient and wettability of oral mucosal tissue: changes induced by a salivary layer. Colloids Surfaces A Physicochem Eng Aspects 2006;276:155–161.

31 Prinz JF, de Wijk RA, Huntjens L: Load dependency of the coefficient of friction of oral mucosa. Food Hydrocolloids 2007;21:402–408.

32 Gans RF, Watson GE, Tabak LA: A new assessment in vitro of human salivary lubrication using a compliant substrate. Arch Oral Biol 1990;35:487–492.

33 Reeh ES, Aguirre A, Sakaguchi RL, Rudney JD, Levine MJ, Douglas WH: Hard tissue lubrication by salivary fluids. Clin Mater 1990;6:151–162.

34 Badawi H, Major P: Three-dimensional orthodontic force measurements. Am J Orthod Dentofacial Orthop 2010;137:299–300.

35 Veeregowda DH, van der Mei HC, de Vries J, Rutland MW, Valle-Delgado JJ, Sharma PK, et al: Boundary lubrication by brushed salivary conditioning films and their degree of glycosylation. Clin Oral Invest 2012;16:1499–1506.

36 Veeregowda DH, van der Mei HC, Busscher HJ, Sharma PK: Influence of fluoride-detergent combinations on the visco-elasticity of adsorbed salivary protein films. Eur J Oral Sci 2011;119:21–26.

37 Lee S, Muller M, Rezwan K, Spencer ND: Porcine gastric mucin (PGM) at the water/poly(dimethylsiloxane) (PDMS) interface: influence of pH and ionic strength on its conformation, adsorption, and aqueous lubrication properties. Langmuir 2005;21:8344–8353.

38 Hatton MN, Loomis RE, Levine MJ, Tabak LA: Masticatory lubrication – the role of carbohydrate in the lubricating property of a salivary glycoprotein albumin complex. Biochem J 1985;230:817–820.

39 Yakubov GE, Mccoll J, Bongaerts JHH, Ramsden JJ: Viscous boundary lubrication of hydrophobic surfaces by mucin. Langmuir 2009;25:2313–2321.

40 Douglas WH, Reeh ES, Ramasubbu N, Raj PA, Bhandary KK, Levine MJ: Statherin – a major boundary lubricant of human saliva. Biochem Biophys Res Communic 1991;180:91–97.

41 Berg CH, Lindh L, Arnebrant T: Intra-oral lubrication of PRP-1, statherin and mucin as studied by AFM. Biofouling 2004;20:65–70.

42 Harvey NM, Carpenter GH, Proctor GB, Klein J: Normal and frictional interactions of purified human statherin adsorbed on molecularly-smooth solid substrata. Biofouling 2011;27:823–835.

43 Macakova L, Yakubov GE, Plunkett MA, Stokes JR: Influence of ionic strength on the tribological properties of pre-adsorbed salivary films. Tribol Int 2011;44:956–962.

44 Veeregowda DH, Busscher HJ, Vissink A, Jager D-J, Sharma PK, van der Mei HC: Role of structure and glycosylation of adsorbed protein films in biolubrication. PloS One 2012;7:e42600.

45 Lundin M, Sandberg T, Caldwell KD, Blomberg E: Comparison of the adsorption kinetics and surface arrangement of 'as received' and purified bovine submaxillary gland mucin (BSM) on hydrophilic surfaces. J Colloid Interface Sci 2009;336:30–39.

46 Zhang YF, Zheng J, Zheng L, Shi XY, Qian LM, Zhou ZR: Effect of adsorption time on the lubricating properties of the salivary pellicle on human tooth enamel. Wear 2013;301:300–307.

47 Macakova L, Yakubov GE, Plunkett MA, Stokes JR: Influence of ionic strength changes on the structure of pre-adsorbed salivary films. A response of a natural multi-component layer. Colloids Surfaces B Biointerfaces 2010;77:31–39.

48 Hannig M, Khanafer AK, Hoth-Hannig W, Al-Marrawi F, Acil Y: Transmission electron microscopy comparison of methods for collecting in situ formed enamel pellicle. Clin Oral Invest 2005;9:30–37.

49 Santos O, Lindh L, Halthur T, Arnebrant T: Adsorption from saliva to silica and hydroxyapatite surfaces and elution of salivary films by SDS and delmopinol. Biofouling 2010;26:697–710.

50 Bradway SD, Bergey EJ, Jones PC, Levine MJ: Oral mucosal pellicle – adsorption and transpeptidation of salivary components to buccal epithelial-cells. Biochem J 1989;261:887–896.

51 Humphrey SP, Williamson RT: A review of saliva: normal composition, flow, and function. J Prosthet Dentistry 2001;85:162–169.

52 Gibbins HL, Carpenter GH: Alternative mechanisms of astringency – what is the role of saliva? J Texture Stud 2013;44:364.

53 Larsen MJ, Jensen AF, Madsen DM, Pearce EIF: Individual variations of pH, buffer capacity, and concentrations of calcium and phosphate in unstimulated whole saliva. Arch Oral Biol 1999;44: 111–117.

54 Gibbins HL, Proctor GB, Yakubov GE, Wilson S, Carpenter GH: Concentration of salivary protective proteins within the bound oral mucosal pellicle. Oral Dis 2013, Epub ahead of print.

55 Yao Y, Lamkin MS, Oppenheim FG: Pellicle precursor proteins: acidic proline-rich proteins, statherin, and histatins, and their crosslinking reaction by oral transglutaminase. J Dent Res 1999;78:1696–1703.

56 Cardenas M, Elofsson U, Lindh L: Salivary mucin MUC5B could be an important component of in vitro pellicles of human saliva: an in situ ellipsometry and atomic force microscopy study. Biomacromolecules 2007;8:1149–1156.

57 Coles JM, Chang DP, Zauscher S: Molecular mechanisms of aqueous boundary lubrication by mucinous glycoproteins. Curr Opin Colloid Interface Sci 2010;15:406–416.

58 Yao Y, Lamkin MS, Oppenheim FG: Pellicle precursor protein crosslinking: Characterization of an adduct between acidic proline-rich protein (PRP-1) and statherin generated by transglutaminase. J Dent Res 2000;79:930–938.

59 Zimmerman JN, Custodio W, Hatibovic-Kofman S, Lee YH, Xiao Y, Siqueira WL: Proteome and peptidome of human acquired enamel pellicle on deciduous teeth. Int J Mol Sci 2013;14:920–934.

60 Siqueira WL, Custodio W, McDonald EE: New insights into the composition and functions of the acquired enamel pellicle. J Dent Res 2012;91:1110–1118.

61 Svendsen IE, Lindh L: The composition of enamel salivary films is different from the ones formed on dental materials. Biofouling 2009;25:255–261.

62 Soares RV, Siqueira CC, Bruno LS, Oppenheim FG, Offner GD, Troxler RF: MG2 and lactoferrin form a heteropic complex in salivary secretions. J Dent Res 2003;82:471–475.

63 Svendsen IE, Lindh L, Elofsson U, Arnebrant T: Studies on the exchange of early pellicle proteins by mucin and whole saliva. J Colloid Interface Sci 2008;321: 52–59.

64 Lendenmann U, Grogan J, Oppenheim FG: Saliva and dental pellicle – a review. Adv Dent Res 2000;14:22–28.

65 Pramanik R, Osailan SM, Challacombe SJ, Urquhart D, Proctor GB: Protein and mucin retention on oral mucosal surfaces in dry mouth patients. Eur J Oral Sci 2010;118:245–253.

66 Punyadeera C, Nouwens AS, Stokes JR, Yakubov G, Cooper-White J: Elucidation of changes in human saliva protein profiles during resting and stimulated states. In: Human Proteome World Congress, Sydney, Australia 2010.

Gleb E. Yakubov
School of Chemical Engineering
University of Queensland
St. Lucia, QLD 4072 (Australia)
E-Mail g.yakubov@uq.edu.au

Ligtenberg AJM, Veerman ECI (eds): Saliva: Secretion and Functions.
Monogr Oral Sci. Basel, Karger, 2014, vol 24, pp 88–98 (DOI: 10.1159/000358791)

Saliva Diagnostics: Utilizing Oral Fluids to Determine Health Status

Christopher A. Schafer[a] · Jason J. Schafer[b] · Maha Yakob[a] ·
Patricia Lima[a] · Paulo Camargo[a] · David T.W. Wong[a]

[a]UCLA Center for Oral/Head and Neck Oncology Research, School of Dentistry, UCLA, Los Angeles, Calif.,
and [b]Department of Pharmacy Practice, Jefferson School of Pharmacy, Thomas Jefferson University,
Philadelphia, Pa., USA

Abstract

Imagine a time where your health status could be available to you without the pain, discomfort and inconvenience of a physical examination. Distant vision of an inconceivable future or impending reality with potentially immeasurable impact? Recent advancements in the field of molecular diagnostics indicate this is not only possible, but closer than we think. Novel discoveries and substantial advancements have revealed that saliva may contain real-time information describing our overall physiological condition. Researchers are now reporting that, like blood and tissue biopsies, oral fluids could be a source of biochemical data capable of detecting certain diseases. What is even more intriguing is that this phenomenon not only applies to local disorders like oral cancer and Sjögren's syndrome, but distant pathologies like autoimmune, cardiovascular and metabolic diseases as well as viral/bacterial infections and even some cancers. These revelations have provided a foundation for the burgeoning field of salivary diagnostics and hence spurred the onset of investigations poised at deciphering the salivary milieu. This paper overviews salivary diagnostics from biomarker development to the multitude of techniques utilized in identifying saliva-based molecular indicators of disease. In doing so, we present oral fluids as an easily accessible noninvasive alternative to traditional diagnostic avenues and not just an essential component of the digestive process. Determining saliva as a credible means of evaluating health status represents a considerable leap forward in health care, one that could lead to enormous translational advantages and significant clinical opportunities.

© 2014 S. Karger AG, Basel

Molecular diagnostics is defined as the application of molecular biology techniques for the purpose of evaluating tissues and biofluids to diagnose, monitor and prognosticate disease. These techniques precisely target molecular and/or microbial entities (also called biomarkers) commonly considered as signs of specific pathologies. Identifying biomarkers at early stages of disease can expedite therapeutic interventions leading to increased survival rates, decreased suffering and low likelihood of recurrence. Current diagnostic assessments require the procurement of testing materials obtained by blood draws, biopsies or other painfully invasive procedures resulting in substantial patient discomfort and augmented health care costs. Furthermore, many disorders do not warrant diagnostic evaluations as a great deal are asymptomatic and typically remain undiagnosed until they have reached an advanced stage when damage is irreversible and treatment futile. Altogether, these statements point to the necessity for noninvasive modalities capable of distinguishing biomarkers indicative of early stage pathogenesis.

To that end, one common ambition among clinicians and basic scientists is to create a credible manner of painless predictive diagnostics. One approach is the exploration of biofluids containing analytes sensitive to our overall health status.

While primarily considered an indispensable element of early digestion, saliva is actually a heterogeneous biofluid comprised of numerous molecules along with a diverse range of microbes. Interestingly, in recent years it has become evident that these very salivary constituents become detectably altered in response to certain disease states. Even so, what is most impressive is that salivary biomarkers not only arise in correlation with oral disorders, but also those of distal tissues and organs. This suggests that oral fluids may represent a substantial reservoir of molecular and microbial information capable of communicating the onset or presence of disease throughout the body.

In consideration, a growing number of studies are focused on elucidating and developing saliva-based biomarkers indicative of both local and systemic diseases. Establishing these markers as credible could usher in a new era of patient evaluation by permitting individuals to avoid the anxiety of traditional diagnostic procedures. The potential impact of utilizing saliva as a means to ascertain the existence or likelihood of disease could markedly advance medicine, as we know it.

We begin our discussion by exploring biomarker development and how disease-specific analytes are scrutinized prior to widespread acceptance and Food and Drug Administration (FDA) approval. Next, we delve into the vast arena of molecular and microbial analyses by describing a number of diagnostic foci. Termed 'salivaomics' these categories include: proteomics, epigenomics, metabolomics, immunomics, microbiome and transcriptomics. Highlighting each library as crucial to biomarker development, we review a selection of their respective biomarkers that lend credence to the continued pursuit of salivary diagnostics. Last we discuss evidence that details one possible mechanism responsible for the induction of systemic disease indicators in salivary secretions.

The discovery of discriminatory biomarkers in oral fluids epitomizes an almost incomparable breakthrough in clinical and translational science. Continuation in this area could credential saliva as an acceptable diagnostic medium, an end point with the potential to bring about a paradigm shift in the discipline of molecular diagnostics.

Biomarker Development

Challenges

In the past decade, the field of molecular diagnostics has made profound leaps forward in biomarker discovery and the development of clinically applicable diagnostic techniques. However, definitive diagnoses can be difficult to achieve based solely on the detection of biomarkers and usually require affirmation by the employment of more traditional and often painful clinical proce-

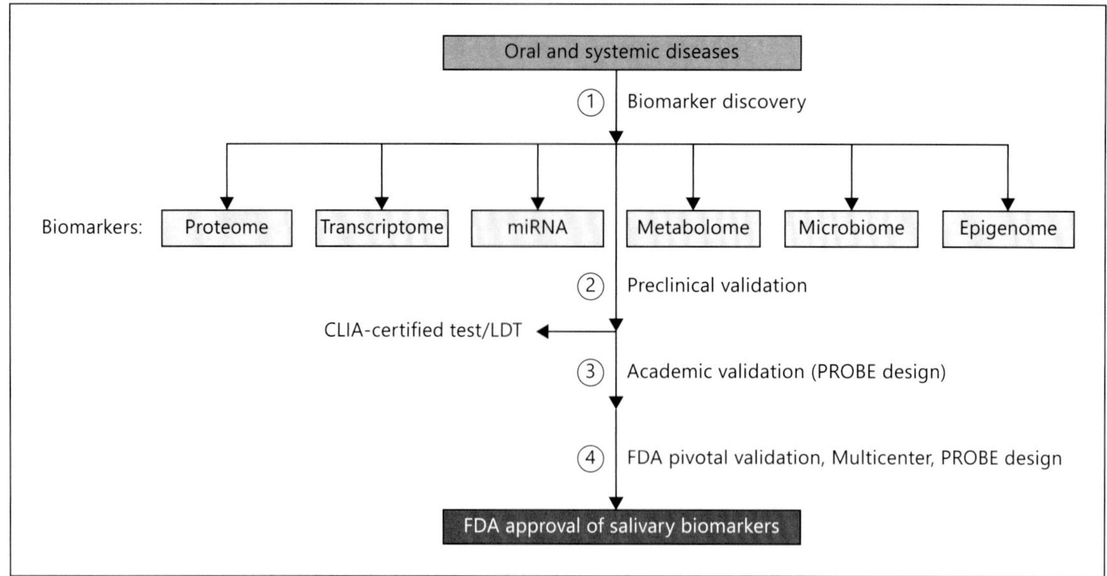

Fig. 1. Process of biomarker discovery through definitive validation and approval. Biomarkers of oral and systemic diseases are discovered using one or more of the 'omics' libraries. Verified biomarkers are subjected to increasing scrutiny and larger independent cohorts until they reach a pivotal multicenter validation. CLIA = Clinical laboratory improvement amendments; LDT = laboratory developed test.

Table 1. Main issues and questions

Challenges	Questions
(1) Identifying molecular biomarkers of disease	Can we identify disease-specific molecules in biofluids?
(2) Establishing noninvasive means of sample collection	Can we devise procedures to procure biofluids while minimizing patient discomfort?
(3) Developing cost-effective platforms for disease detection	Can we create affordable, accurate and efficient modes of sample analysis?

dures. These obstacles, among other limitations, challenge diagnosticians in their ongoing quest to establish reliable markers of disease. Table 1 lists a summary of the main issues confronting us along with the questions they pose.

Process
The process of biomarker development is a difficult task with multiple levels of evaluation prior to definitive approval by the FDA (fig. 1). During the initial stages samples are analyzed using 1 or more of the 6 'omics' libraries: proteome, transcriptome, immunome, metabolome, microbiome and epigenome. Each library contains a vast collection of information potentially useful in determining an individual's current state of health. Following discovery, verified biomarkers proceed through successive levels of validation using prospective randomized open blinded end point (PROBE) designed studies and indepen-

Schafer · Schafer · Yakob · Lima · Camargo · Wong

dent cohorts [1]. FDA approval is achieved after substantial evaluations have been performed using large patient populations and advanced statistical algorithms.

Salivaomics

Proteomics

In 2004 the National Institute of Dental and Craniofacial Research provided funding to comprehensively decipher, catalogue and annotate the human salivary proteome. These efforts revealed the salivary proteome as a sizeable collection of up to 1,166 diverse proteins [2, 3], the majority of which are synthesized and subsequently secreted into the oral cavity by the acinar cells of the salivary gland. This suggests that saliva-based proteomic constituents are a product of the salivary glands, which, in turn, may be subject to both proximate and external factors. Consequently, the salivary proteome could be indicative of both local and distant diseases.

To evaluate the capacity of salivary proteomic entities as diagnostic metrics, scientists frequently employ traditional laboratory techniques including: liquid chromatography, gel and capillary electrophoresis, nuclear magnetic resonance, mass spectrography, immunoassays and lectin probe analysis [4]. However, more contemporary methods like 2-dimensional gel electrophoresis coupled with mass spectrometry are also utilized and have allowed investigators to analyze hundreds of salivary analytes [3, 5].

Although its proteomic content is estimated to be only 30% that of blood [6], saliva is actively being investigated as a rich source of protein biomarkers capable of discerning healthy from diseased subjects. In fact, researchers have reported unique proteomic profiles indicative of oral cancer, cystic fibrosis, endocrinological and connective tissue disorders, dental caries, periodontitis and even breast carcinoma [7–9].

Transcriptomics

Composed of coding (messenger RNA) and noncoding RNAs (small nucleolar RNAs, microRNAs, i.e. miRNAs, and small nuclear RNAs), the transcriptome makes a significant source of potentially relevant diagnostic information. miRNAs are small (19–21 nucleotides) regulatory molecules that functionally interact with and potentially silence the translation of complementary antisense RNA sequences [10]. Similar to messenger RNAs, miRNAs can be aberrantly expressed in irregularly functioning or tumorigenic cell lines [11]. These attributes suggest that transcriptomic evaluation may reveal highly specific discriminatory indicators.

Interestingly whole saliva as well as saliva supernatant harbor both miRNAs and total RNA. In support of this statement, investigators probing the salivary transcriptome have identified over 1,000 miRNAs and more than 3,000 species of messenger RNAs [12]. Suggesting that oral fluids contain viable biochemical data, these results support the notion that saliva is an informative biofluid that may prove useful in discerning health and disease. With this in mind, investigations at the UCLA School of Dentistry have discovered and are pursuing the identification of salivary biomarkers for oral, pancreatic, lung and gastric cancers along with Sjögren's syndrome, diabetes and periodontal disease [12–15].

Although gene chip arrays and quantitative polymerase chain reaction denote the most common methodologies for generating diagnostic values, the development of recent technologies promises to expand our evaluative capacity. For example, the utilization of RNA-sequencing platforms allows for the quantitation of differentially represented transcripts in a solution [16]. Employing this and other global high throughput methods will permit a more thorough understanding of the salivary transcriptome and its role in diagnostics.

Methylomics

Epigenetics is the study of heritable changes in gene expression that are not reflected by changes in DNA sequence. While several epigenetic mechanisms including DNA methylation, histone modification and RNA interference have been implicated in disease pathogenesis, none have been studied more than the methylome [17].

Known to play a crucial role in cellular differentiation, DNA methylation allows cells to maintain unique characteristics by controlling and modulating gene expression [18]. Methylation patterns are typically established during embryogenesis and subsequently remain stable in normal tissues [19]. However, hypermethylation of CpG islands has been shown to be carcinogenic [20]. Identifying individuals with hypermethylated promoters may expedite the initiation of therapies and reveal the efficacy of screening high-risk populations.

Remarkably, it is not uncommon to find DNA within salivary secretions. Hence, several investigations have focused their efforts on elucidating patterns of promoter hypermethylation in salivary DNA. One such study discovered gene panels capable of discerning oral squamous cell carcinoma patients from controls with 62–77% sensitivity and 83–100% specificity [21]. Additionally, two separate groups revealed regions of promoter hypermethylation in the saliva of head and neck squamous cell carcinoma patients [22]. These reports suggest that saliva-based DNA may potentially be an efficacious analyte for the detection and surveillance of head and neck and oral squamous cell carcinomas.

In addition, a number of studies have explored the effect of both local and global epigenetic alterations with regard to age [23]. For instance, a recent study found a significant association between age and the methylation status of 80 genes in salivary samples [24]. Continued and future analyses in this area could serve as a forerunner for the establishment of saliva-based predictive screenings for age-related disease.

Remarkably, salivary methylomic analysis techniques may also have a role in forensic science. Preliminary studies indicate that bisulfite sequencing of DNA from saliva and other biofluids could successfully discriminate 5 tissue-specific differentially methylated regions [25]. Altogether these results highlight the potential impact of utilizing saliva-based DNA as not only an indicator of disease, but also a discriminator of tissues. Concertedly, the attributes discussed here may not only prove valuable in terms of discerning certain disease states, but also in assessing critical biological evidence for legal procedures.

Immunomics

Perhaps the most significant advancement in the field of salivary diagnostics is the identification of immunological markers of systemic infections. Suspected to enter saliva from the blood through absorption and subsequent secretion by the salivary glands, these include the antigens, antibodies and nucleic acids of infecting pathogens. Intriguingly, this assumed exchange of blood-based molecules into our salivary glands results in comparable concentrations of immunological markers, a characteristic allowing for the noninvasive diagnosis and/or monitoring of systemic infections via oral fluids. In understanding this, the FDA has recently approved ELISA kits designed to probe saliva for indicators of HIV-1 and -2 [26]. Although further testing is required to confirm diagnosis, these statements represent a substantial step towards establishing saliva as a diagnostic medium.

While AIDS is at the forefront of immunological salivary diagnostics, significant progress has also been noted with regard to hepatitis A, B and C whose antigens and antibodies are routinely detected in the saliva of infected individuals. Even though hepatitis B and hepatitis C diagnosis and monitoring are primarily blood based, researchers are now reporting quantifiable antibodies and antigens in saliva [27, 28]. Accordingly, commercially available tests capable of identifying saliva-based hepatitis C virus antibodies are under development. Although the accuracy of the results obtained from this test draw parallels with serum immunoassays (97.5%), it has yet to obtain FDA approval. Along with this, recent investigations describe the efficacy of utilizing saliva to analyze hepatitis A virus immunizations and evaluate acute exposures using IgM, IgA, IgG and hepatitis A virus RNA tests.

In considering the simplicity of saliva collection and its potential as a diagnostic medium, oral fluids have rapidly become the focus of investigation for several worldwide endemic microbes. Among these are malaria, dengue, Ebola and *Mycobacterium tuberculosis* as well as herpes simplex virus, Epstein-Barr virus, cytomegalovirus and human herpes virus. Utilizing saliva to detect these disorders could potentially decrease their associated mortality rates. However, continued research is necessary to expand beyond the now commercially available HIV and hepatitis C virus tests to include a wider variety of microbes.

Microbiome
Microorganisms found in the human oral cavity have been referred to as the oral microflora, the oral microbiota and more recently the oral microbiome [29]. While certain estimates state that approximately 500–700 bacterial species commonly inhabit the oral cavity, others suggest that it may be as high as 10,000 [30]. Not surprisingly, studies have shown that alterations in the oral microbial community have been correlated with the development of oral pathologies, malodor and dental caries [7–9]. Though fascinating, what poses a greater question here is the relationship connecting the oral microbiome and systemic disorders.

Despite the lack of clarity describing how oral bacteria and distal diseases are linked, clinicians and basic scientists continue to explore this phenomenon. As a result oral microbial candidate biomarkers are now being identified for Crohn's disease, chronic pancreatitis, pancreatic cancer and even obesity [31].

While thought provoking these studies have yet to be definitively validated and recognized as credible. Continued efforts in this area could distinguish the oral microbiome as a reservoir of microbial constituents with the ability to identify both local and systemic diseases, a truly unexpected but exceedingly important feature of the oral cavity.

Metabolomics
In general a metabolome is composed of a variety of small molecules including: amino acids, organic acids, peptides, lipids, nucleic acids, etc. Metabolomics is the science of searching for unique chemical fingerprints of these molecules within specific biological samples [32]. Identifying discriminatory metabolomic alterations could not only aid in determining disease diagnosis, but also facilitate the selection, design and modification of care plans.

In addition to tissues and cells, a number of biological fluids including saliva contain metabolomic analytes [33]. As such, salivary metabolomics is an emerging and promising field that offers another unique perspective on the use of oral fluids in molecular diagnostics. In fact, researchers are now reporting significant discrepancies in several classes of metabolites (dipeptides, amino acids, carbohydrates, lipids and nucleotides) when comparing those afflicted with periodontitis to healthy subjects [34]. Similar studies analyzing the metabolic profiles discerned individuals with oral, pancreatic or breast cancers from their respective healthy controls [35–37]. From this it can be deduced that disease-specific metabolomic signatures may be embedded in salivary secretions. Continued work in this area could lead to the identification and employment of additional discriminatory metabolites useful in medical screening as well as disease diagnosis and monitoring.

Although the application of metabolomics is still in the preliminary stages, its influence continues to grow each year. Integrating the knowledge obtained from different 'omics' studies may help in understanding the etiology of unique metabolomic profiles as well as the mechanism of how individuals respond to certain therapies [38].

The concurrent utilization of these 'omics' libraries will be the canon of the future. This multitiered strategy will provide numerous layers of data revealing a clearer and, perhaps, more concise picture of a patient's disease state or potential response to treatments. The additive effect of these techniques may not only further our understanding of disease pathophysiology, but also result in more favorable clinical outcomes.

Induction of Salivary Biomarkers

The revelation that oral fluids are comprised of analytes capable of reflecting health status simultaneously presents tremendous translation potential and a unique biological question, namely how biomarkers come to exist in saliva at all. In other words, what are the cellular and molecular mechanisms that drive the induction of salivary indicators for systemic disorders? With this in mind a number of researchers are now attempting to reveal the finer details describing the communicative interplay between salivary glands and distant tissues.

Intercellular communication is a fundamental and imperative process where individual cells molecularly correspond with neighboring and distant environments in an effort to coordinate physiological activity within multicellular organisms [39, 40]. Classically this phenomenon is considered to occur locally by means of direct cell contacts and remotely through juxtacrine, paracrine and endocrinological signaling modalities [41]. Typically enzymes, hormones, cytokines and chemokines exemplify these long-range protein-based mechanisms. Each method is well studied and commonly results in a mea-

surable cellular response in target cells [41]. The very nature of these processes and their consequential effects emphasizes the regulatory complexity of the body and illustrates how tissues convey and react to information delivered via extracellular pathways. Although significant efforts have been applied to analyzing proteins in this capacity, recent studies have revealed a role for nucleic acids as major contributors to the communicative interaction of remote cell populations [42–44].

Despite the ubiquity of ribonucleases it is not uncommon to find RNA in the serum, plasma, breast milk, urine and saliva of healthy and diseased patients [45–47]. Suggesting that secreted RNA can perpetuate in the extracellular milieu, these findings call into question the biological significance of these molecules in this context. Alternatively, what role do these molecules serve outside their respective host cells?

In considering this question one must understand that RNA innately possesses unique physicochemical properties that enable it to facilitate a number of biological pathways [48, 49]. As a product of transcription, both coding and noncoding RNAs efficiently store genetic information, a characteristic that highlights their potential impact on cellular dynamics. These attributes, in compilation with its surprising extracellular stability, provide credence for the recent emergence of RNA as a possible mediator of long-distance extracellular communication [48, 49]. Including RNA as a potential effector of intercellular crosstalk adds an exciting and previously hidden dimension of molecular interaction and genetic regulation within a biological system [48]. However, its well-known propensity for degradation draws attention to the need for a protective vehicle that not only serves to shield the RNA from enzymatic destruction, but also as a means for transport and delivery. Accordingly, studies have shown that certain microvesicular bodies called exosomes fulfill these parameters by encasing and delivering function-

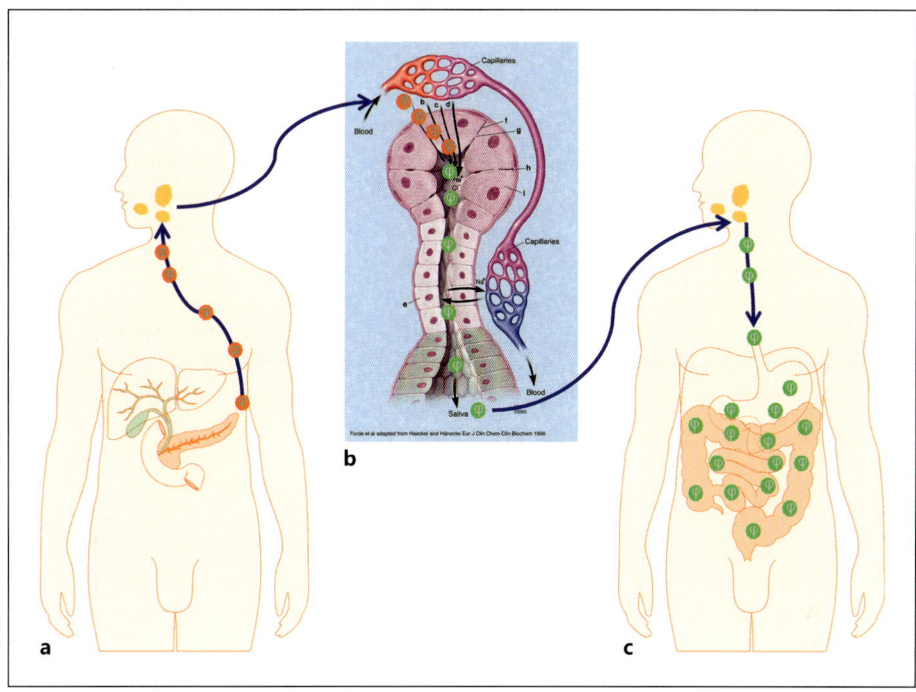

Fig. 2. Mechanistic induction of salivary biomarkers. **a** Tissue-specific exosomes shed from distant pathologies (e.g. pancreatic cancer) circulate throughout the vasculature. **b** Disease-derived exosomes (orange) are taken up by salivary gland parenchymal cells and subsequently released to saliva as conduits of systemic disease biomarkers (green). **c** When swallowed, disease-specific salivary exosomes may exert a systemic effect on the host via mucosal immunity in the gastrointestinal tract.

al genetic information from donor to recipient cells [50].

Measuring 30–100 nm in diameter, exosomes are small secretory lipidic assemblies formed by the invagination of intracellular vesicles. Exosomes are randomly secreted and routinely found in the conditioned media of multiple cell types as well as a number of biological fluids ranging from blood and saliva to breast milk and urine [40, 44, 51, 52].

Exosomes reportedly contain viable tissue-specific biochemical cargo that upon delivery has been shown to induce unique molecular activity within target cells [52–54]. To that end researchers are now showing that exosomal constituents may possess the capacity to manipulate immune responses, angiogenesis, coagulation and numerous other complex physiological mechanisms [55, 56] in recipient tissues. Concertedly, these statements put forth the idea that exosomes may distinctively function as a means of protecting and shuttling biologically active information for the purposes of regulating cell communication [50].

In consideration, it could be inferred that circulating exosomes derived from diseased tissues could interact with and potentially deliver a unique message to salivary glands (fig. 2). This event may alter the molecular constituency of saliva by modifying biochemical activity within saliva-secreting cells inducing them to generate and release a discriminatory biomarker signature [57]. Although the mechanistic minutiae of such an occurrence have yet to be explained, this exosomal interplay could account for the onset of disease-specific saliva-based biomarkers. While much remains to be proven in this regard, furthering our understanding of the processes that yield these outcomes will not only help define the role of exosomes in extracellular communication, but also potentially influence the advance of novel strategies for disease diagnostics, monitoring and therapeutics.

Saliva: Diagnostic Medium of the Future

The noninvasive nature of saliva collection in collusion with high throughput analytical technologies highlights the efficacy of pursuing saliva as a diagnostic medium. Although the realization of salivary diagnostics as a ubiquitously accepted standard of analysis is yet to be realized, its clinical and commercial potential is immeasurable. The imminent emergence of new technologies along with the establishment of validated biomarker profiles will propel saliva to the forefront of diagnostics and set the pace for what is to come.

Disclosure Statement

David Wong is cofounder of RNAmeTRIX Inc., a molecular diagnostic company. He holds equity in RNAmeTRIX and serves as a company Director and Scientific Advisor. The University of California also holds equity in RNAmeTRIX. Intellectual property that David Wong invented and which was patented by the University of California has been licensed to RNAmeTRIX. Additionally, he is a paid consultant to PeriRx.

References

1 Pepe MS, Feng Z, Janes H, Bossuyt PM, Potter JD: Pivotal evaluation of the accuracy of a biomarker used for classification or prediction: standards for study design. J Natl Cancer Ins 2008;100: 1432–1438.

2 Denny P, Hagen FK, Hardt M, Liao L, Yan W, Arellanno M, Bassilian S, Bedi GS, Boontheung P, Cociorva D, Delahunty CM, Denny T, Dunsmore J, Faull KF, Gilligan J, Gonzalez-Begne M, Halgand F, Hall SC, Han X, Henson B, Hewel J, Hu S, Jeffrey S, Jiang J, Loo JA, Ogorzalek Loo RR, Malamud D, Melvin JE, Miroshnychenko O, Navazesh M, Niles R, Park SK, Prakobphol A, Ramachandran P, Richert M, Robinson S, Sondej M, Souda P, Sullivan MA, Takashima J, Than S, Wang J, Whitelegge JP, Witkowska HE, Wolinsky L, Xie Y, Xu T, Yu W, Ytterberg J, Wong DT, Yates JR 3rd, Fisher SJ: The proteomes of human parotid and submandibular/ sublingual gland salivas collected as the ductal secretions. J Proteome Res 2008; 7:1994–2006.

3 Hu S, Loo JA, Wong DT: Human Saliva Proteome Analysis. Ann NY Acad Sci 2007;1098:323–329.

4 Kawas SA, Rahim ZHA, Ferguson DB: Potential uses of human salivary protein and peptide analysis in the diagnosis of disease. Arch Oral Biol 2012;57:1–9.

5 Hu S, Xie Y, Ramachandran P, Ogorzalek Loo RR, Li Y, Loo JA, Wong DT: Large-scale identification of proteins in human salivary proteome by liquid chromatography/mass spectrometry and two-dimensional gel electrophoresis-mass spectrometry. Proteomics 2005;5:1714–1728.

6 Schulz BL, Cooper-White J, Punyadeera CK: Saliva proteome research: current status and future outlook. Crit Rev Biotechnol 2013;33:246–259.

7 Paster BJ, Dewhirst FE: Molecular microbial diagnosis. Periodontol 2000 2009;51:38–44.

8 Takeshita T, Suzuki N, Nakano Y, Shimazaki Y, Yoneda M, Hirofuji T, Yamashita Y: Relationship between oral malodor and the global composition of indigenous bacterial populations in saliva. Appl Environm Microbiol 2010;76: 2806–2814.

9 Yang F, Zeng X, Ning K, Liu K-L, Lo C-C, Wang W, Chen J, Wang D, Huang R, Chang X, Chain PS, Xie G, Ling J, Xu J: Saliva microbiomes distinguish cariesactive from healthy human populations. ISME J 2012;6:1–10.

10 Lee RC, Feinbaum RL, Ambros V: The *C. elegans* heterochronic gene lin-4 encodes small RNAs with antisense complementarity to lin-14. Cell 1993;75: 843–854.

11 Croce CM: Causes and consequences of microRNA dysregulation in cancer. Nat Rev Genet 2009;10:704–714.

12 Brinkmann O, Wong DT: Salivary transcriptome biomarkers in oral squamous cell cancer detection. Adv Clin Chem 2011;55:21–34.

13 Hu S, Vissink A, Arellano M, Roozendaal C, Zhou H, Kallenberg CGM, Wong DT: Identification of autoantibody biomarkers for primary Sjögren's syndrome using protein microarrays. Proteomics 2011;11:1499–1507.

14 Lee H, Park M, Choi A, An J, Yang W, Shin K-J: Potential forensic application of DNA methylation profiling to body fluid identification. Int J Legal Med 2012;126:55–62.

15 Zhang L, Farrell JJ, Zhou H, Elashoff D, Akin D, Park NH, Chia D, Wong D: Salivary transcriptomic biomarkers for detection of resectable pancreatic cancer. Gastroenterology 2010;138:949–957.

16 Spielmann N, Ilsley D, Gu J, Lea K, Brockman J, Heater S, Setterquist R, Wong DTW: The human salivary RNA transcriptome revealed by massively parallel sequencing. Clin Chem 2012;58: 1314–1321.

17 Vasilatou D, Papageorgiou S, Dimitriadis G, Pappa V: Epigenetic alterations and microRNAs: new players in the pathogenesis of myelodysplastic syndromes. Epigenetics 2013;8:561–570.

18 Ohgane J, Yagi S, Shiota K: Epigenetics: the DNA methylation profile of tissue-dependent and differentially methylated regions in cells. Placenta 2008; 29(suppl):29–35.

19 Grønbæk K, Hother C, Jones PA: Epigenetic changes in cancer. APMIS 2007; 115:1039–1059.

20 Herman JG, Baylin SB: Gene silencing in cancer in association with promoter hypermethylation. N Engl J Med 2003; 349:2042–2054.

21 Viet CT, Schmidt BL: Methylation array analysis of preoperative and postoperative saliva DNA in oral cancer patients. Cancer Epidemiol Biomarkers Prev 2008;17:3603–3611.

22 Carvalho AL, Henrique R, Jeronimo C, Nayak CS, Reddy AN, Hoque MO, Chang S, Brait M, Jiang W-W, Kim MM, Claybourne Q, Goldenberg D, Khan Z, Khan T, Westra WH, Sidransky D, Koch W, Califano JA: Detection of promoter hypermethylation in salivary rinses as a biomarker for head and neck squamous cell carcinoma surveillance. Clin Cancer Res 2011;17:4782–4789.

23 Boks MP, Derks EM, Weisenberger DJ, Strengman E, Janson E, Sommer IE, Kahn RS, Ophoff RA: The relationship of DNA methylation with age, gender and genotype in twins and healthy controls. PloS One 2009;4:e6767.

24 Bocklandt S, Lin W, Sehl ME, Sánchez F, Sinsheimer JS, Horvath S, Vilain E: Epigenetic predictor of age. PloS One 2011; 6:e14821.

25 Lee Y-H, Kim J, Zhou H, Kim B, Wong D: Salivary transcriptomic biomarkers for detection of ovarian cancer: for serous papillary adenocarcinoma. J Mol Med 2012;90:427–434.

26 United States Food and Drug Administration: First rapid home-use HIV kit approved for self-testing. FDA Consumer Health Information 2012. http://www.fda.gov/downloads/ForConsumers/ConsumerUpdates/UCM311690.pdf (accessed May 6, 2013).

27 Chen L, Liu F, Fan X, Gao J, Chen N, Wong T, Wu J, Wen SW: Detection of hepatitis B surface antigen, hepatitis B core antigen, and hepatitis B virus DNA in parotid tissues. Int J Infect Dis 2009; 13:20–23.

28 Gonzalez V, Martro E, Folch C, Esteve A, Matas L, Montoliu A, Grifols JR, Bolao F, Tural C, Muga R, Parry JV, Ausina V, Casabona J: Detection of hepatitis C virus antibodies in oral fluid specimens for prevalence studies. Eur J Clin Microbiol Infect Dis 2008;27:121–126.

29 Dewhirst FE, Chen T, Izard J, Paster BJ, Tanner ACR, Yu W-H, Lakshmanan A, Wade WG: The human oral microbiome. J Bacteriol 2010;192:5002–5017.

30 Keijser BJF, Zaura E, Huse SM, van der Vossen JMBM, Schuren FHJ, Montijn RC, ten Cate JM, Crielaard W: Pyrosequencing analysis of the oral microflora of healthy adults. J Dent Res 2008;87: 1016–1020.

31 Farrell JJ, Zhang L, Zhou H, Chia D, Elashoff D, Akin D, Paster BJ, Joshipura K, Wong DT: Variations of oral microbiota are associated with pancreatic diseases including pancreatic cancer. Gut 2012;61:582–588.

32 Jordan KW, Nordenstam J, Lauwers GY, Rothenberger DA, Alavi K, Garwood M, Cheng LL: Metabolomic characterization of human rectal adenocarcinoma with intact tissue magnetic resonance spectroscopy. Dis Colon Rectum 2009; 52:520–525

33 Walsh MC, Brennan L, Malthouse JPG, Roche HM, Gibney MJ: Effect of acute dietary standardization on the urinary, plasma, and salivary metabolomic profiles of healthy humans. Am J Clin Nutr 2006;84:531–539.

34 Barnes VM, Ciancio SG, Shibly O, Xu T, Devizio W, Trivedi HM, Guo L, Jönsson TJ: Metabolomics reveals elevated macromolecular degradation in periodontal disease. J Dent Res 2011;90:1293–1297.

35 Sugimoto M, Wong D, Hirayama A, Soga T, Tomita M: Capillary electrophoresis mass spectrometry-based saliva metabolomics identified oral, breast and pancreatic cancer-specific profiles. Metabolomics 2010;6:78–95.

36 Wei J, Xie G, Zhou Z, Shi P, Qiu Y, Zheng X, Chen T, Su M, Zhao A, Jia W: Salivary metabolite signatures of oral cancer and leukoplakia. Int J Cancer 2011;129:2207–2217.

37 Yan S-K, Wei B-J, Lin Z-Y, Yang Y, Zhou Z-T, Zhang W-D: A metabonomic approach to the diagnosis of oral squamous cell carcinoma, oral lichen planus and oral leukoplakia. Oral Oncol 2008; 44:477–483.

38 Wang L: Pharmacogenomics: a systems approach. Wiley Interdiscip Rev Syst Biol Med 2010;2:3–22.

39 Chen X, Liang H, Zhang J, Zen K, Zhang C-Y: Secreted microRNAs: a new form of intercellular communication. Trends Cell Biol 2012;22:125–132.

40 Thery C, Ostrowski M, Segura E: Membrane vesicles as conveyors of immune responses. Nat Rev Immunol 2009;9: 581–593.

41 Pant S, Hilton H, Burczynski ME: The multifaceted exosome: BIogenesis, role in normal and aberrant cellular function, and frontiers for pharmacological and biomarker opportunities. Biochem Pharmacol 2012;83:1484–1494.

42 Bellingham SA, Guo B, Coleman B, Hill AF: Exosomes: vehicles for the transfer of toxic proteins associated with neurodegenerative diseases? Front Physiol 2012;3:124.

43 Hood JL, San RS, Wickline SA: Exosomes released by melanoma cells prepare sentinel lymph nodes for tumor metastasis. Cancer Res 2011;71:3792–3801.

44 Yang C, Robbins PD: The roles of tumor-derived exosomes in cancer pathogenesis. Clin Dev Immunol 2011;2011: 11.

45 Fleischhacker M, Schmidt B: Circulating nucleic acids (CNAs) and cancer – a survey. Biochim Biophys Acta 2007; 1775:181–232.

46 Kosaka N, Iguchi H, Yoshioka Y, Takeshita F, Matsuki Y, Ochiya T: Secretory mechanisms and intercellular transfer of microRNAs in living cells. J Biol Chem 2010;285:17442–17452.

47 O'Driscoll L: Extracellular nucleic acids and their potential as diagnostic, prognostic and predictive biomarkers. Anticancer Res 2007;27:1257–1265.

48 Dinger ME, Mercer TR, Mattick JS: RNAs as extracellular signaling molecules. J Mol Endocrinol 2008;40:151–159.

49 Scott WG: Ribozymes. Curr Opin Struct Biol 2007;17:280–286.
50 Lässer C, Eldh M, Lötvall J: Isolation and characterization of RNA-containing exosomes. J Vis Exp 2012;59:e3037.
51 Record M, Subra C, Silvente-Poirot S, Poirot M: Exosomes as intercellular signalosomes and pharmacological effectors. Biochem Pharmacol 2011;81:1171–1182.
52 Vlassov AV, Magdaleno S, Setterquist R, Conrad R: Exosomes: current knowledge of their composition, biological functions, and diagnostic and therapeutic potentials. Biochim Biophys Acta 2012;1820:940–948.
53 Dimov I, Jankovic Velickovic L, Stefanovic V: Urinary exosomes. Sci Wrld J 2009;9:1107–1118.
54 Thery C, Zitvogel L, Amigorena S: Exosomes: composition, biogenesis and function. Nat Rev Immunol 2002;2:569–579.
55 Lakkaraju A, Rodriguez-Boulan E: Itinerant exosomes: emerging roles in cell and tissue polarity. Trends Cell Biol 2008;18:199–209.
56 Pegtel DM, van de Garde MDB, Middeldorp JM: Viral miRNAs exploiting the endosomal/exosomal pathway for intercellular cross-talk and immune evasion. Biochim Biophys Acta 2011;1809:715–721.
57 Lau CS, Wong DTW: Breast cancer exosome-like microvesicles and salivary gland cells interplay alters salivary gland cell-derived exosome-like microvesicles in vitro. PLoS One 2012;7:e33037.

David T.W. Wong
UCLA Oral/Head and Neck Oncology Research Center
UCLA School of Dentistry
10833 Le Conte Avenue, CHS 73-034
Los Angeles, CA 90095 (USA)
E-Mail dtww@ucla.edu

Schafer · Schafer · Yakob · Lima · Camargo · Wong

Ligtenberg AJM, Veerman ECI (eds): Saliva: Secretion and Functions.
Monogr Oral Sci. Basel, Karger, 2014, vol 24, pp 99–108 (DOI: 10.1159/000358864)

The Use of Saliva Markers in Psychobiology: Mechanisms and Methods

Jos A. Bosch

Department of Clinical Psychology, University of Amsterdam, Amsterdam, The Netherlands

Abstract

In the social sciences, the use of saliva parameters has greatly expanded in recent years from the measurement of steroid hormones, like cortisol, and now includes a wide range of biochemical parameters. These salivary constituents can be broadly classified into two groups: (1) constituents that enter saliva from plasma (e.g. hormones, inflammatory markers, drug chemicals) and (2) constituents that are produced locally by the saliva glands (e.g. α-amylase, secretory IgA). Reliable measurement of blood-borne constituents assumes a constant saliva/plasma ratio (SPR), which implies that the concentration in saliva truthfully follows intra- and interindividual variations in plasma. The first part of this review discusses the main determinants of the SPR: the mechanism by which plasma constituents enter saliva (i.e. passive diffusion, active transport, ultrafiltration, leakage) and associated physiochemical factors. The second part of this review provides an overview of central and peripheral neural mechanisms that regulate saliva gland function and the release of glandular proteins. This section provides a neurobiological underpinning for a section, which addresses methodological implications for the as-sessment of glandular secretions. Salivary psychophysiology is a fast-growing field and the time seems ripe for more rigorous methodological studies that may help this discipline to reach its full potential.

The use of salivary parameters in psychobiological research, and especially in stress research, seems to be an obvious choice. Everyone is familiar with the experience of a dry mouth during public speaking, and emotional inhibition of salivary flow is the basis of what possibly is the oldest use of saliva to monitor a psychological phenomenon: the Rice Test [1]. This ancient lie detection test required an accused to take a mouthful of dry rice. If anxiety, and presumably guilt, inhibited salivation to such an extent that the defendant could not form an adequate bolus for chewing and swallowing, then punishment would follow. One of the first systematic scientific accounts on the use of saliva in psychological studies were reviewed by Karl Lashley, who also invented a device for the collection of parotid saliva that still carries his name [2]. His research focused on the study of conditioning, and attempted to replicate Pavlov's reflex conditioning in humans. These attempts failed: humans show no conditioning of salivary reflexes. This is mainly because they show little of a salivation reflex in response to olfactory stimuli [3], a fact that is so counterintuitive that it remains mentioned in textbooks in spite of systematic evidence of the contrary [3].

Contemporary uses of saliva in psychological research have shifted their focus from salivary flow rate to the measurement of salivary constituents, such as hormones (e.g. cortisol, testosterone, dehydroepiandrosterone, i.e. DHEA, melatonin), drugs and their metabolites (e.g. cotinine, amphetamines), immunological molecules (e.g. secretory immunoglobulin A, i.e. IgA, C-reactive protein) and saliva gland secretory proteins (e.g. α-amylase). On the basis of their origin from which they enter saliva, these salivary constituents can broadly be classified into two groups: (1) constituents that enter saliva from plasma and serum and (2) constituents that are produced locally by the saliva glands. This classification is relevant to methodology and interpretation of salivary findings, as each source involves different mechanisms. The present paper will address these mechanisms and associated methodological issues. Specifically, the first section will address the question how plasma constituents enter saliva, and the second part will address how the secretion of saliva and salivary gland proteins is regulated. That section is also devoted to methodological implications of these different mechanisms for saliva collection.

Mechanisms

How Do Plasma Constituents Get into Saliva?
Saliva is not a filtrate of the blood, and most plasma constituents can enter saliva only by diffusion or by an active transport mechanism [4]. To be able to reliably monitor plasma constituents in saliva there should be a constant saliva/plasma ratio (SPR), which implies that the concentration in saliva truthfully follows intra- and interindividual variations in plasma [5]. The glucocorticoid hormone cortisol is a good example of a plasma constituent that has an excellent SPR, which significantly contributes to its usefulness as a psychophysiological readout. This section will address the biochemical factors that determine a constant SPR, which clarifies why cortisol is somewhat of an exception and that most plasma-derived molecules cannot be reliably assessed in saliva.

Passive Diffusion
Passive diffusion is thought to be the major mechanism by which steroid hormones enter saliva. A plasma molecule that enters saliva by diffusion has to cross 5 barriers: the capillary wall; the interstitial space; the basal cell membrane of the acinus cell or duct cell; the cytoplasm of the acinus or duct cell, and the luminal cell membrane [6]. A primary physicochemical factor that determines the efficiency of this diffusion must therefore be the ability to pass through cell membranes; hence lipid solubility is a key determinant of an optimal SPR, whereby apolar lipophilic molecules more readily pass through cell membranes than lipophobic molecules [5]. Another molecular determinant that is conducive to membrane permeation is small molecular size (smaller molecules diffuse more readily). Again taking cortisol as the prototypical example: this molecule is apolar, hence lipid soluble, and very small (MW 362). Several other steroid hormones have comparable properties, such as testosterone and DHEA, and these properties likewise explain why their salivary concentrations reliably reflect those found in plasma [7]. The above properties are also shared by a number of drugs and drug metabolites, and have for that reason been of use in drug monitoring studies. Two examples are cotinine and theophylline [for a comprehensive overview, see 4].

A factor that hinders diffusion into saliva, and thus is associated with a low SPR and poor inferences about plasma levels, is the fact that many plasma constituents are bound to plasma proteins. Examples are albumin, selective soluble receptors, and designated carrier molecules. For example, about 95% of plasma cortisol is bound to albumin or corticoid-binding globulin [4]. Proteins typically lack the properties for an optimal SPR as described above, and bound molecules will therefore not be able to diffuse into saliva. However, it has been argued that this factor may actually pose a blessing in disguise, as protein-bound plasma molecules tend to be biologically inactive [7]. Thus, capturing the unbound, or 'free', fraction of a plasma molecule provides information on biologically active concentrations [8].

Active Transport, Ultrafiltration and Leakage
Other mechanisms by which molecules may enter saliva are active transport, ultrafiltration and via leaky patches such as sites of tissue damage and inflammation. Active transport is somewhat of an exception. The best-known example is the transport of IgA, an immunological protein that is produced and secreted by local B cells. Secreted IgA is subsequently picked up by the poly-IgA receptor (also designated as secretory component) expressed on glandular cells and the secretory component-IgA complex is then transported through these cells and secreted into saliva as secretory IgA [9]. Ion pumps in the cell membrane are another example of an active transport mechanism. It has been shown that such transport is under control of the autonomic nervous system (see section further below). There is no known mechanism for active transport of plasma hormones into saliva [4].

Ultrafiltration, the third mechanism, involves the seeping-through of plasma molecules via the spaces between acinus and ductal cells and the tight junctions between cells of secretory units. This mode of transportation requires that the molecules are relatively small (up to 2 kDa). Components that enter saliva via ultrafiltration have concentrations that are typically several hundred times lower than in plasma [6]. Sulfated steroids like DHEA sulfate and estriol sulfates, which cannot passively diffuse through the cell membrane because of their polarity, are believed to enter saliva via this route [4, 7]. The efficiency of transport via ultrafiltration is likely to be variable within and between individuals, although this has not been a topic of systematic research. For example, it has been shown that tight junctions may widen under elevated autonomic activation of the salivary glands [10]. Molecules that enter saliva via ultrafiltration are also affected by the diluting effects of the saliva flow rate.

Finally, plasma proteins such as cytokines, protein hormones or acute phase proteins are too large to enter saliva via either diffusion or ultrafiltration. These molecules enter saliva via leaky patches, such as sites of inflammation and abrasions, and via the crevicular fluid (a plasma filtrate emanating from the junction between the oral mucosa and dentine). Transudation of plasma compounds from the oral mucosa, a form of passive transport (largely) through osmotic (oncotic) pressure, is another route by which such molecules may be transported to saliva [4, 6]. Taken together, the saliva concentrations of plasma proteins would therefore seem the most difficult to determine reliably, and susceptible to intraindividual variations related to local (as opposed to systemic) factors such as local inflammation, tissue damage and oral health.

In summary, this section reviewed 4 main mechanisms by which plasma constituents can enter saliva. It makes clear why cortisol is such an ideal molecule to assess in saliva; being small, apolar and having a low propensity to ionize, it readily diffuses through cells and tissues. On the other hand, it may become clear why data on immune proteins in saliva, like C-reactive protein and cytokines, have to be interpreted with some caution [11, 12]. Such proteins enter saliva via routes that are highly variable, both between and within participants, and the efficiency in which they enter saliva is likely determined by local factors hereby undermining their validity as markers of systemic processes and states.

How Do Glandular Proteins Become Secreted into Saliva?

As outlined in the introductory paragraph, a second class of salivary constituents that are routinely measured in the social sciences are the products of saliva glands. Mostly these are proteins, although ions like sodium and potassium have been utilized in psychological studies as well [10]. The salivary protein that has received the most attention is salivary α-amylase (sAA), and this section will thus be particularly relevant to this readout [13, 14]. sAA is a digestive enzyme that breaks down insoluble starch into soluble maltose and dextrin. It can be measured quickly and reproducibly with commercially available kits based on chromogenic or fluorogenic substrates that utilize its enzymatic activity. This activity is expressed in units per milliliter and taken as a proxy for sAA concentration [15]. sAA has become popular as a means to monitor sympathetic activity. This application seems consistent with known glandular biology in man and in animal models, described in detail below, whereby the parasympathetic nerves are mainly responsible for the secretion of 'high fluid-low protein saliva' (hence are said to control fluid secretion), and the sympathetic nerves mainly elicit secretion of 'low fluid-high protein saliva' (hence are said to regulate salivary protein secretion) including the secretion of sAA [16]. The sections that follow below will provide a review of the neural regulation of glandular function. It presents the available data on CNS centers involved in coordinating saliva gland activity, which supports the validity of using salivary parameters in stress research. It then presents information on the local regulation of saliva

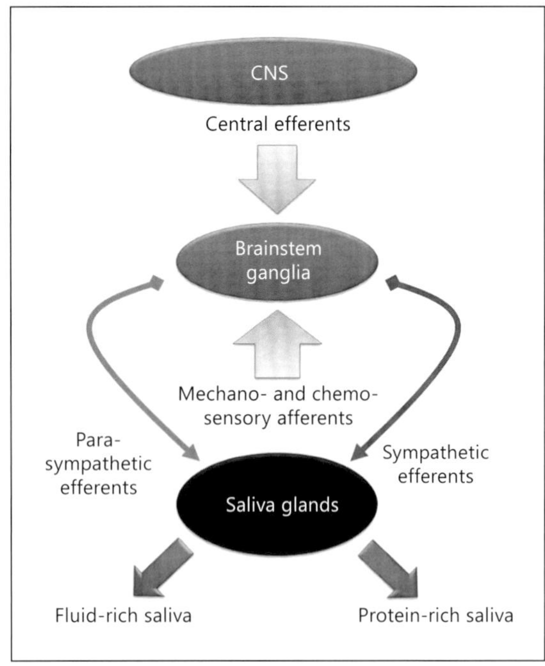

Fig. 1. Glandular activity is regulated by the autonomic ganglia in the brainstem, which receive inputs from local sensory nerves (eliciting reflex secretion) as well as from signals emanating from the CNS. Local reflex activation (e.g. by chewing) thus competes with central regulating signals (e.g. affective states).

gland function and discusses that protein secretion involves both branches of the autonomic nervous system. The information presented below can be seen to provide a neurobiological underpinning of a recent review [17] that specifically addressed methodological and interpretational issues related to the use of sAA in the social sciences.

Autonomic Control of Salivary Gland Activity
Central Regulation of Glandular Function
The saliva gland receives direct inputs from so-called primary salivatory ganglia in the brainstem (parasympathetic) and upper thoracic spinal cord (sympathetic), which regulate reflex-like responses in response to chewing and possibly taste [18, 19] (fig. 1). The hypothesis that the salivary gland

function is susceptible to modulation by emotional states must therefore imply that these primary centers receive inputs from higher neural structures. Historically, two types of studies have provided support for this assumption. The first type comprises studies in which specific brain areas are stimulated and subsequent bodily changes are recorded. Brown [20] cites animal research showing that stimulation of the dorsomedial and ventromedial hypothalamic nuclei increases salivation, whereas stimulation of the lateral and posterior hypothalamus inhibits salivation. These limbic nuclei primarily regulate basic motivational states such as hunger, thirst, fear, anger and thermoregulation. Stimulation studies also indicated that changes in salivation are part of a constellation of responses typical for stressful states. For example, Wang [21] cites evidence showing that cats in which the ventromedial part of the hypothalamus is stimulated display behavioral responses indicative of intense anxiety which, paradoxically, are accompanied by increases in salivation.

The second type of supporting data comes from histological studies in which central connections are identified by tracing the retrograde axonal transport of the enzyme horseradish peroxidase or a virus. The former method traces the neurons that directly synapse on a specific ganglion (in this case the salivatory nuclei), whereas the latter method also traces second and higher-order synapses. The horseradish peroxidase studies show that the primary parasympathetic salivary centers receive *direct* inputs from various cell groups in the mid- and forebrain [22]. Major forebrain centers include the paraventricular and lateral hypothalamus, the central nucleus of the amygdala and the bed nucleus of the stria terminalis. All of these have been heavily implicated in fear (e.g. amygdala) and anxiety (e.g. stria terminalis) and the generation of coordinated stress responses [19].

Central sympathetic regulation is less well determined but includes the paraventricular and lateral hypothalamus, central gray matter and rostral ventrolateral medulla. The central gray matter is also an important relay station for coordinating fight-flight responses, whereas the rostral ventrolateral medulla is well recognized for its role in the integration of cardiovascular and respiratory reflexes [19, 22].

Local Autonomic Connections
The aforementioned CNS networks innervate local ganglia that are primarily responsible for regulating glandular activity [16] (fig. 1). Parasympathetic control of the submandibular and sublingual glands originates in the inferior salivatory nucleus, which is situated in the pons, medial to the front (rostral) part of the solitary nucleus. The axons of these neurons follow the facial nerve (7th cranial nerve, or chorda tympani), where they directly or indirectly, via the submandibular ganglion, synapse in the glands. Parasympathetic control of the parotid gland is governed by the superior salivatory nucleus, which is situated in the caudal part of the medulla (in humans the exact position has not yet been identified). The efferent nerves enter the periphery via the glossopharyngeal nerve (the 9th cranial nerve), terminating on the parotid gland via the otic ganglion. The precise location of the sympathetic salivary center has not been identified either. However, it is known that the sympathetic preganglionic neurons in the upper thoracic nerves (T_1–T_5) connect with the salivary glands [19, 20, 22].

Salivary Gland Response Specificity Further Regulated on a Local Cellular Level
Considering the multiple convergences of synaptic inputs from many brain loci, and the potential of sympathetic-parasympathetic interaction, one may speculate that the glands are capable of generating highly diverse patterns of responses. For example, in humans it is shown that the salivary glands are capable of producing highly differentiated protein responses to differ-

ent stressors [23]. Matsuo et al. [24] showed that within the rat submandibular gland, and after cutting the sympathetic nerves, different forms of reflex-induced secretion (e.g. induced by heat or chewing) produce different protein patterns. Thus, differential responses can already be achieved with only one functional autonomic branch in one gland. Such a differentiation is possible because the axons of one type (sympathetic or parasympathetic) in a gland do not contain the same transmitter substances to the same extent [25–27]. For example, in the human submandibular gland the number of autonomic fibers that are positive for neuropeptide Y, vasointestinal peptide or galanine is significantly higher around the mucous acini than around the serous acini [16]. Furthermore, different sets of autonomic transmitters are released with different rates and patterns of neuronal activation [16]. Whereas release of the conventional transmitters acetylcholine and noradrenaline generally occurs with every propagated nerve impulse, neuropeptides are primarily released with higher frequencies of nerve stimulation, particularly when these impulses are applied in bursts [27, 28]. Thus, the autonomic nerves seem capable of utilizing Morse-like semantics of differential transmitter release. It is also appreciated that neurotransmitter release does not occur from all potential effector sites along an axon with each impulse, which is presumably depending on the local interaction with other neural outputs [25, 26]. Thus, even at the level of a single axon a propagated impulse can generate events that differ from one synaptic location to another. Finally, both between glands and between cells within glands, there are marked differences in the density and patterning of receptors that are responsive to the various messenger substances conveyed by the autonomic nerves. These involve both classic autonomic receptor types, which respond to either noradrenaline or acetylcholine, and the nonadrenergic-noncholinergic receptors that respond to other autonomic messenger

substances, such as peptides, nitric oxide and purines [16]. Differential activation of these receptor types can cause a myriad of additive, synergistic or antagonistic intracellular responses, ultimately resulting in a protein release that is capable of being differentially regulated both between and within glands.

In summary, the salivary glands form a highly sophisticated end point in the CNS control of local immune defenses, capable of responding instantly and with a high level of specificity to potential sources of harm (e.g. food, stress, inflammation). This remarkable ability, together with their strategic location at the portal of entry to the respiratory and gastrointestinal tract, make these glands ideally suited to provide the host with a first line of defense. However, as another side of the coin, this complexity hinders a straightforward interpretation of glandular output in terms of sympathetic or parasympathetic activity. The extent to which such complexity actually determines the data is of course ultimately an empirical question, and it is therefore regrettable that systematic research is largely absent from the psychophysiological literature.

Methods

Stimulated or Unstimulated Saliva Collection?
Stimulation of oral mechanoreceptors (e.g. during chewing) or chemoreceptors (e.g. during application of citric acid) induces local autonomic reflex activity involving the local ganglia that were described above [16]. This reflex mechanism has implications for the interpretation of plasma constituents that enter saliva by processes other than passive diffusion (see above) as well as for the interpretation of saliva gland proteins such as sAA [17]. Essentially, increasing saliva flow rate will cause further dilution of the plasma products that enter saliva by active transport, ultrafiltration or other leakage routes. This

may for example be the case with salivary C-reactive protein, an acute phase protein secreted by the liver in states of elevated inflammation, which may explain the low correspondence between saliva and plasma levels in some studies. Such variability will add to the already existing noise, as discussed, as such molecules will have a poor SPR.

More pertinent is the activation of reflex secretion to the interpretation of saliva gland proteins such as sAA. Reflex secretion is a mechanism of glandular secretion that is independent of central regulation (e.g. stress) and thus overrules the central neural effects of stress [29] (fig. 1). A familiar and everyday example of this phenomenon is that the sense of dry mouth during stress, due to a parasympathetic inhibition, can easily be overruled by simply chewing on gum. This situation would be comparable to one whereby the effects of psychological stress on heart rate are studied in individuals who are engaged in strenuous physical activity: the physiological needs for optimal blood flow and gas exchange, regulated by lower brain centers and local reflexes, will take priority over any higher cognitive processes, such as the heart rate deceleration during an orienting response or the heart rate acceleration during a defense response. For that reason cardiovascular psychophysiologists test participants under conditions of minimal or no movement [30]. Similarly, in stress studies saliva should be collected without engaging any local reflex mechanisms [17].

The lack of appreciation for the fundamental distinction between unstimulated and stimulated saliva is expressed in the unstandardized manner in which saliva is collected. For example, participants are instructed 'to gently move the Salivette around in the mouth' or 'to chew on the Salivette' [17]. In this way the researcher will collect data that are open to all sources of confounding. As argued elsewhere [17], one may wonder whether a relaxed study participant will chew with the same vigor as a distressed participant, or whether

the instruction to 'gently move the Salivette around' is acted upon with the same gentleness by female and male participants (or children vs. adults, etc.). The strength of mechanical stimulation corresponds to the amount of saliva produced [31] and thus affects fluid levels and protein secretion independently of central regulation by affective states.

In fact, and discussed in more detail elsewhere [17], even with standardization there are decreases in sAA output and concentration from individual glands within the first several minutes of chewing-induced secretion [32]. Regretfully, the duration of saliva collection is typically not standardized.

A final problem associated with engaging local reflex stimulation of the saliva glands is that it will drastically change salivary protein composition. This effect is again independent of the higher CNS regulation of saliva gland function which the researcher is aiming to capture. This confounding is due to two characteristics of the saliva glands that will interact. First, the different saliva glands vary in their responsivity to reflex activation, e.g. by chewing. Second, the different saliva glands differ in their protein composition, e.g. the parotid gland and palatinal glands are very rich in sAA, whereas sublingual and submandibular gland secretions are much lower in sAA. It has been estimated that without any mechanical or gustatory stimulation (i.e. passive) most saliva is secreted by the submandibular glands and that only about 20% derives from the parotid gland [33, 34]. However, during chewing the contributions change and about half of all saliva now derives from the parotid glands. Because parotid saliva contains a 4- to 10-fold higher sAA concentration than submandibular saliva [35], sAA concentrations in whole saliva are likely to drastically change independently of higher CNS regulation. While this caveat has been noted [17], its impact on the interpretation of sAA data has to receive further empirical confirmation.

Measuring Concentration or Secretion?

The assessment of secretion rates is relevant to quantitatively interpret salivary protein levels, e.g. sAA, in terms of autonomic nervous system activity [17]. Most saliva proteins that are synthesized and secreted by the salivary glands are released upon activation of the innervating sympathetic nerves (some proteins, such as mucins, may be released upon parasympathetic activation only [23]). The amount of protein that is secreted per unit of time thus corresponds to the level of local sympathetic activation [16]. This implies that protein output per unit of time would be more adequate as a proxy for sympathetic activity. Furthermore, salivary protein concentration reflects the combined effect of salivary flow rate (which is largely parasympathetic) and protein secretion (which, in sympathetically innervated glands, is largely sympathetic). Thus, if proteins like sAA are to be regarded as measures of sympathetic activity, it would seem pertinent that the parasympathetic effect on salivary flow rate is adjusted. This issue not so much reflects an empirical issue but rather a mathematical one, because [protein secretion (U/min) = protein concentration (U/ml) × salivary flow rate (ml/min)].

Use of Collection Devices

A final methodological note is on the use of absorbent materials, such as cotton or polyester sponges (e.g. Salivette). Their use is often implemented because it is thought that participants feel uncomfortable with spitting into a tube, although this idea does not appear to be empirically supported: the only study on this issue thus far observed equal unpleasantness and uneasiness during spitting versus the use of Salivettes [36]. Notwithstanding, use of absorbent materials would seem practical in situations and populations where saliva collection by spitting or drooling might be difficult to perform (e.g. newborns, exercise studies). The researcher should make a balanced decision on whether disadvantages outweigh potential benefits. It is for example well es-

tablished that the Salivette, being one of the most commonly used collection devices, can introduce substantial measurement error to salivary analytes [36–39]. In general absorbent materials not fully release their contents, and this retention can show a strong inverse relation with the amount of fluid absorbed [17]. For example, Salivettes did not release any of the absorbed sAA when the cotton contained 0.25 ml of saliva, which approximates the normal amount of unstimulated saliva produced over 1 min, and only partly released sAA with higher volumes, thus generating a substantial underestimation of true values [37]. In contrast, some analytes rather show elevated values, such as DHEA and testosterone. A further limitation is that salivary flow rate (see section above) cannot be reliably assessed using absorbent materials due to a ceiling/saturation effect (see discussion in Beltzer et al. [39]). It has therefore been recommended that researchers use established standardized procedures for unstimulated whole saliva collection, as described elsewhere [40, 41].

Conclusion

Few methodological papers have been published in psychophysiology and allied fields that specifically address mechanisms of saliva secretion and methods for analyses, and researchers typically utilize methods that were originally validated for cortisol research. The first section of this paper discussed that the transport of cortisol into saliva involves mechanisms that do not apply to most other salivary constituents, and that reliable assessment of these components may therefore require other methods. On that note, a recent review of the stress literature showed that most sAA studies (but similar observations may apply to research on other saliva constituents) do not control for the potentially confounding effects of salivary flow rate, do not standardize saliva collection in terms of stimulation or collection duration, and by default use absorbent materials that are known to distort sAA values [17]. Further, many reports do not provide details on how saliva was collected or how participants were instructed, which seem to imply that researchers (and reviewers) did not think this was a critical issue. We hope that the information in this paper may compel readers to revise this perception, and spur further methodological research in this area.

References

1 Mandel ID: Salivary diagnosis: promises, promises. Ann NY Acad Sci 1993; 694:1–10.
2 Lashley KS: The human salivary reflex and its use in psychology. Psychol Rev 1916;23:446–464.
3 Lee VM, Linden RW: An olfactory-parotid salivary reflex in humans? Exp Physiol 1991;76:347–355.
4 Kaufman E, Lamster IB: The diagnostic applications of saliva – a review. Crit Rev Oral Biol Med 2002;13:197–212.
5 Jusko WJ, Milsap RL: Pharmacokinetic principles of drug distribution in saliva. Ann NY Acad Sci 1993;694:36–47.
6 Pfaffe T, Cooper-White J, Beyerlein P, Kostner K, Punyadeera C: Diagnostic potential of saliva: current state and future applications. Clin Chem 2011;57: 675–687.
7 Lewis JG: Steroid analysis in saliva: an overview. Clin Biochem Rev 2006;27: 139–146.
8 Kudielka BM, Gierens A, Hellhammer DH, Wust S, Schlotz W: Salivary cortisol in ambulatory assessment – some dos, some don'ts, and some open questions. Psychosom Med 2012;74:418–431.
9 Brandtzaeg P, Farstad IN, Johansen FE, Morton HC, Norderhaug IN, Yamanaka T: The B-cell system of human mucosae and exocrine glands. Immunol Rev 1999;171:45–87.
10 Bosch JA, Ring C, de Geus EJC, Veerman ECI, Amerongen AV: Stress and secretory immunity. Int Rev Neurobiol 2002; 52:213–253.
11 Pace TW, Negi LT, Dodson-Lavelle B, Ozawa-de Silva B, Reddy SD, Cole SP, Danese A, Craighead LW, Raison CL: Engagement with cognitively-based compassion training is associated with reduced salivary C-reactive protein from before to after training in foster care program adolescents. Psychoneuroendocrinology 2013;38:294–299.

12 Byrne ML, O'Brien-Simpson NM, Reynolds EC, Walsh KA, Laughton K, Waloszek JM, Woods MJ, Trinder J, Allen NB: Acute phase protein and cytokine levels in serum and saliva: a comparison of detectable levels and correlations in a depressed and healthy adolescent sample. Brain Behav Immun 2013;34:164–175.
13 Nater UM, Rohleder N: Salivary alpha-amylase as a non-invasive biomarker for the sympathetic nervous system: current state of research. Psychoneuroendocrinology 2009;34:486–496.
14 Rohleder N, Nater UM: Determinants of salivary alpha-amylase in humans and methodological considerations. Psychoneuroendocrinology 2009;34:469–485.
15 Mandel AL, Peyrot des Gachons C, Plank KL, Alarcon S, Breslin PA: Individual differences in AMY1 gene copy number, salivary alpha-amylase levels, and the perception of oral starch. PLoS One 2010;5:e13352.
16 Proctor GB, Carpenter GH: Regulation of salivary gland function by autonomic nerves. Auton Neurosci 2007;133:3–18.
17 Bosch JA, Veerman EC, de Geus EJ, Proctor GB: α-Amylase as a reliable and convenient measure of sympathetic activity: don't start salivating just yet! Psychoneuroendocrinology 2011;36:449–453.
18 Bosch JA, Ring C, de Geus EJ, Veerman EC, Amerongen AV: Stress and secretory immunity. Int Rev Neurobiol 2002; 52:213–253.
19 Bosch JA, Veerman ECI: Neural regulation of the saliva glands during stress and exercise; in Sreebny LM, Vissink A (eds): Dry Mouth: A Clinical Guide. New York, Blackwell-Munksgaard, 2008, pp 213–226.
20 Brown CC: The parotid puzzle: a review of the literature on human salivation and its applications to psychophysiology. Psychophysiology 1970;7:65–85.

21 Wang SC: Central nervous representation of salivary secretion. Monographs in oral Biology 1964;3:145–159.
22 Matsuo R: Central connections for salivary innervations and efferent impulse formation; in Garrett JR, Ekstrom J, Anderson LC (eds): Neural Mechanisms of Salivary Gland Secretion. Basel, Karger, 1999, pp 26–43.
23 Bosch JA, de Geus EJC, Veerman ECI, Hoogstraten J, Nieuw Amerongen AV: Innate secretory immunity in response to laboratory stressors that evoke distinct patterns of cardiac autonomic activity. Psychosom Med 2003;65:245–258.
24 Matsuo R, Garrett JR, Proctor GB, Carpenter GH: Reflex secretion of proteins into submandibular saliva in conscious rats, before and after preganglionic sympathectomy. J Physiol 2000;527:175–184.
25 Garrett JR: Nerves in the main salivary glands; in Garrett JR, Ekstrom J, Anderson LC (eds): Neural Mechanisms of Salivary Gland Secretion. Basel, Karger, 1999, pp 1–25.
26 Garrett JR: Effects of autonomic nerve stimulations on salivary parenchyma and protein secretion; in Garrett JR, Ekstrom J, Anderson LC (eds): Neural Mechanisms of Salivary Gland Secretion. Basel, Karger, 1999, pp 59–79.
27 Ekstrom J: Role of nonadrenergic, noncholinergic autonomic transmitters in salivary glandular activities in vitro; in Garrett JR, Ekstrom J, Anderson LC (eds): Neural Mechanisms of Salivary Gland Secretion. Basel, Karger, 1999, pp 94–130.
28 Ekstrom J, Asztely A, Tobin G: Parasympathetic non-adrenergic, non-cholinergic mechanisms in salivary glands and their role in reflex secretion. Eur J Morphol 1998;36(suppl):208–212.
29 Garrett JR: The proper role of nerves in salivary secretion: a review. J Dent Res 1987;66:387–397.

30 Bosch JA, de Geus EJ, Carroll D, Anane LA, Helmerhorst EJ, Edwards KM: A general enhancement of autonomic and cortisol responses during social evaluative threat. Psychosom Med 2009;71: 877–885.

31 Hector MP, Linden RW: The possible role of periodontal mechanoreceptors in the control of parotid secretion in man. Q J Exp Physiol 1987;72:285–301.

32 Proctor GB, Carpenter GH: Chewing stimulates secretion of human salivary secretory immunoglobulin A. J Dent Res 2001;80:909–913.

33 Humphrey SP, Williamson RT: A review of saliva: normal composition, flow, and function. J Prosthet Dent 2001;85:162–169.

34 Schenkels LCPM, Veerman ECI, Nieuw Amerongen AV: Biochemical composition of human saliva in relation to other mucosal fluids. Crit Rev Oral Biol Med 1995;6:161–175.

35 Veerman EC, van den Keybus PA, Vissink A, Nieuw Amerongen AV: Human glandular salivas: their separate collection and analysis. Eur J Oral Sci 1996; 104:346–352.

36 Strazdins L, Meyerkort S, Brent V, D'Souza RM, Broom DH, Kyd JM: Impact of saliva collection methods on sIgA and cortisol assays and acceptability to participants. J Immunol Methods 2005;307:167–171.

37 DeCaro JA: Methodological considerations in the use of salivary alpha-amylase as a stress marker in field research. Am J Hum Biol 2008;20:617–619.

38 Harmon AG, Towe-Goodman NR, Fortunato CK, Granger DA: Differences in saliva collection location and disparities in baseline and diurnal rhythms of alpha-amylase: a preliminary note of caution. Horm Behav 2008;54:592–596.

39 Beltzer EK, Fortunato CK, Guaderrama MM, Peckins MK, Garramone BM, Granger DA: Salivary flow and alpha-amylase: collection technique, duration, and oral fluid type. Physiol Behav 2010; 101:289–296.

40 Navazesh M: Methods for collecting saliva. Ann NY Acad Sci 1993;694:72–77.

41 Navazesh M, Kumar SK: Measuring salivary flow: challenges and opportunities. J Am Dent Assoc 2008;139(suppl): 35S–40S.

Jos A. Bosch
Department of Clinical Psychology
University of Amsterdam
NL–1018 XA Amsterdam (The Netherlands)
E-Mail j.a.bosch@uva.nl

Ligtenberg AJM, Veerman ECI (eds): Saliva: Secretion and Functions.
Monogr Oral Sci. Basel, Karger, 2014, vol 24, pp 109–125 (DOI: 10.1159/000358792)

Xerostomia

Konstantina Delli[a] · Fred K.L. Spijkervet[a] · Frans G.M. Kroese[b] ·
Hendrika Bootsma[b] · Arjan Vissink[a]

Departments of [a]Oral and Maxillofacial Surgery and [b]Rheumatology and Clinical Immunology,
University of Groningen, University Medical Center Groningen, Groningen, The Netherlands

Abstract

Xerostomia is the subjective feeling of oral dryness. The major causes are Sjögren's syndrome (SS), medication and radiotherapy to the head and neck. SS is a chronic systemic autoimmune disease characterized by infiltration of the exocrine glands, the salivary and lacrimal ones in particular. The pathogenesis involves systemic B cell hyperactivity and T cell lymphocytes targeting glandular epithelial cells. About 7.5% of patients with SS develop malignant B cell lymphoma, mostly mucosa-associated tissue lymphomas. Certain classes of drugs can induce hyposalivation and/or xerostomia by, e.g., targeting neurotransmitters and receptors. As a result, amongst others the production of fluid and electrolytes in salivary glands can be reduced and the salivary composition can change. During head and neck radiotherapy, the administration of high doses to the major salivary glands, which are located in the periphery of the head, leads to progressive loss of glandular function and a diminished salivary output. Reduction of the dose and the volume of irradiated salivary glands by advanced radiotherapy techniques can be highly beneficial for patients.

© 2014 S. Karger AG, Basel

Xerostomia is the subjective feeling of oral dryness. The term is derived from the Greek words *xeros'* (ξηρός), meaning 'dry', and *stoma* (στόμα), meaning 'mouth'. The prevalence of xerostomia is estimated to be between 13 and 63% [1]. It is more prominent in women, in elderly subjects and in individuals housed in long-term care facilities. A number of factors has been associated with transient or persistent xerostomia (table 1). This paper will focus on the three most common causes: Sjögren's syndrome (SS), medication and radiotherapy of the head and neck.

Sjögren's Syndrome

SS was first described by the Swedish ophthalmologist Henrik Sjögren in 1930. It is a chronic inflammatory and lymphoproliferative disorder that is principally characterized by chronic infiltration of the exocrine glands. It commonly affects the salivary and lacrimal glands, resulting in xerostomia and dry eyes (keratoconjunctivitis sicca). These symptoms may be accompanied by extraglandular manifestations, evident in almost any or-

Table 1. Influence of conditions on salivary gland function

	Flow rate	Changed composition	Xerostomia
Xerogenic drugs	↓	+	+
Sjögren's syndrome	↓	+	+
Head and neck radiotherapy	↓	+	+
Chronic inflammatory connective tissue diseases			
Scleroderma	↓	?	+
Mixed connective tissue disease	↓	?	+
Chronic inflammatory bowel diseases			
Crohn's disease	→	+	+
Ulcerative colitis	→	+	−
Celiac disease	→	+	−
Autoimmune liver diseases	↓	?	+
Musculoskeletal disorders			
Fibromyalgia	↓	?	+
Chronic fatigue syndrome	↓	?	+
Amyloidosis	↓	?	+
Endocrine disorders			
Diabetes mellitus	↓	+/−	+
Hyperthyroidism	↑	+	−
Hypothyroidism	↓	?	+
Cushing's syndrome	→	+	−
Addison's disease	→	+	−
Neurological disorders			
CNS trauma	↓	?	?
Cerebral palsy	↓	+	?
Bell's palsy	↓	?	?
Parkinson's disease	↓	+	+
Alzheimer's disease	↓	+	+
Holmes-Adie syndrome	↓	?	+
Burning mouth syndrome	→	+	+
Infectious diseases			
Epidemic parotitis	?	?	?
HIV/AIDS	↓	+/−	+
Hepatitis C virus	↓	?	+
Epstein-Barr virus	?	?	?
Tuberculosis	?	?	?
Local bacterial salivary gland infections	↓	+	?
Genetic disorders			
Salivary gland aplasia	↓	?	?
Cystic fibrosis	↓	+	?
Ectodermal dysplasia	↓	+	−
Prader-Willi syndrome	↓	+	?
Metabolic disturbances			
Water and salt balance	↓	+	+
Sodium retention syndrome	↓	+	+
Malnutrition	↓	+	+
Eating disorders			
Bulimia nervosa	↓	+/−	+
Anorexia nervosa	↓	+	+
Cancer-associated disturbances			
Chemotherapy	↓	+/−	+
Graft-versus-host disease	↓	+	+
Advanced cancer/terminally ill patients	↓	?	+

Modified after Jensen et al. [62], 2010. ↓ = Reduced; ↑ = increased; → = same; + = yes; − = no; ? = unknown.

gan. According to the American-European Consensus Group (AECG) classification criteria [2], SS can be distinguished in primary SS, in case no other autoimmune disease is present, and secondary SS, in case additional connective tissue diseases are present, such as rheumatoid arthritis, systemic lupus erythematosus, scleroderma and mixed connective tissue disease.

Epidemiology

SS is one of the most common rheumatic diseases, with a prevalence of 0.5–1% in the total population. It affects mainly women (i.e. female-to-male ratio equals 9:1). The median age of occurrence is around 50 years, although it can arise in all ages. In rheumatoid arthritis, the prevalence of SS is around 30% and approximately 20% of patients with systemic lupus erythematosus fulfil the criteria for secondary SS.

Clinical Features

Glandular Manifestations

SS primarily affects lacrimal and salivary glands. With respect to the eyes' symptoms, dryness may result in a sensation of itching, grittiness and soreness; nevertheless, the eyes' appearance can be normal. Ocular complaints may include photosensitivity/photophobia, erythema, eye fatigue, decreased visual acuity, discharge from the eyes, and the sensation of a film across the visual field [3] (fig. 1). Progressive keratitis may result in loss of vision in patients with SS. Ocular complications may include corneal ulceration, vascularization, opacification and, rarely, perforation [4].

Another prominent symptom of SS is xerostomia. This symptom is often associated with dysgeusia, difficulty in eating dry food, problems in speaking for long periods of time, burning sensation of the mouth, discomfort while wearing dentures and increased risk of dental caries, especially cervically (fig. 2), and oral infection, in particular candidiasis. At the onset of SS, the mouth appears to be normally moisturized but while the disease progresses, saliva diminishes and becomes foamy,

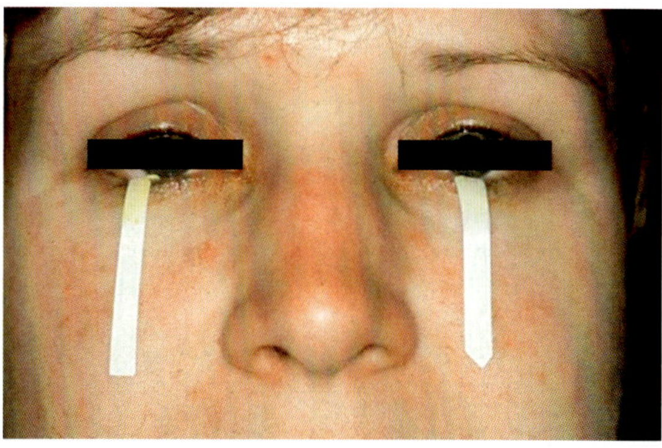

Fig. 1. Dry eye due to diminished tear production, examined with Schirmer's test.

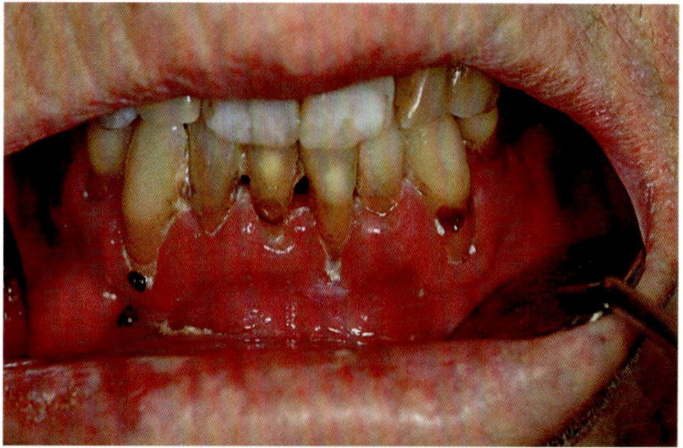

Fig. 2. Tender and dry oral mucosa, characteristically forming fine wrinkles.

lacking the usual pooling in the floor of the mouth. In advanced disease, the oral mucosa is tender and dry and characteristically forms fine wrinkles (fig. 3). The tongue, in particular, often becomes fissured and exhibits atrophy of the papillae.

Enlargement of especially the parotid and submandibular glands (fig. 4) is a common phenomenon in patients with SS. Swelling of the salivary glands is usually bilateral, may be nonpainful to slightly tender and intermittent to persistent in nature. The swelling is generally attributed to the presence of an autoinflammatory process in these glands, and stasis of saliva can result in secondary infection, encouraging further swelling. The development of lymphomas, in most cases in the pa-

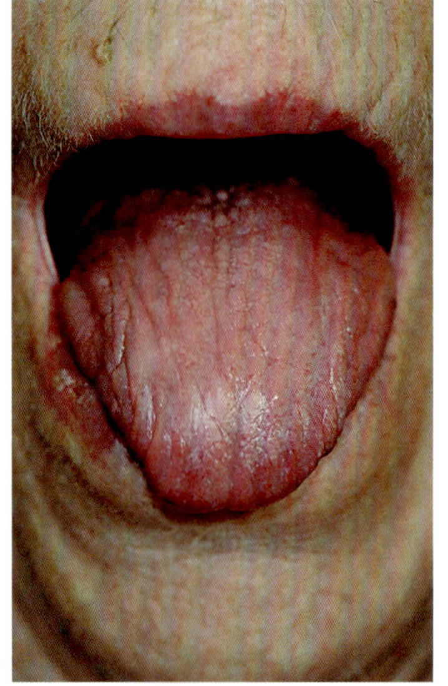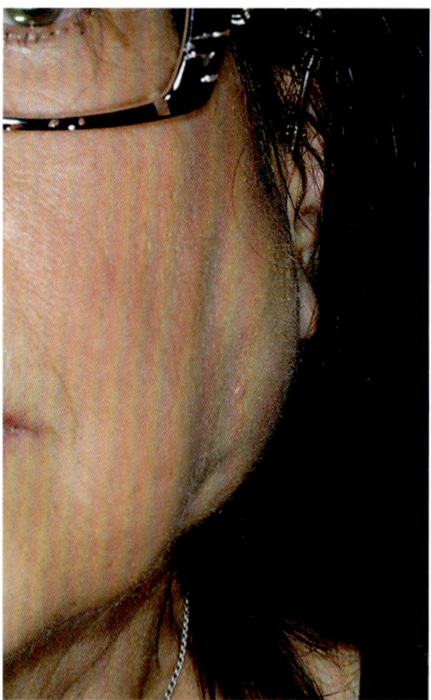

Fig. 3. Red, fissured tongue with atrophy of the papillae.
Fig. 4. Enlarged parotid salivary gland due to the development of MALT lymphoma.

3   **4**

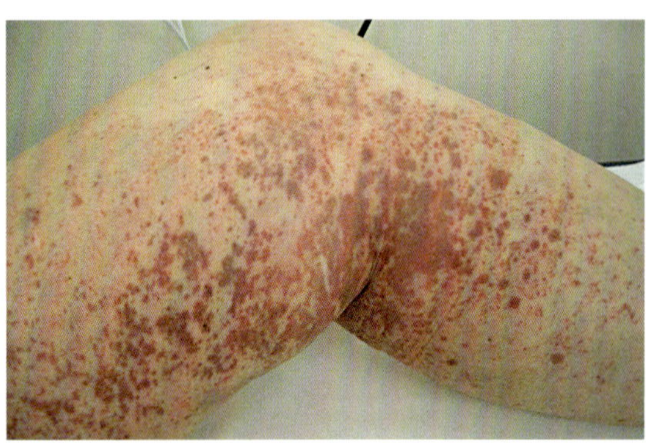

Fig. 5. Purpura is a common extraglandular manifestation in SS patients.

rotid gland, can also lead to more persistent unilateral glandular enlargement. Dryness can occur at the mucosal surfaces of upper and lower respiratory tracts resulting in nonproductive cough [5]. Dry skin affects about 55% of SS patients while in female patients with SS desiccation of the vagina results in dyspareunia [6].

Lymphoma Development

About 7.5% of patients with SS develop malignant B cell lymphoma, 48–75% of which is of the MALT-type (Mucosa Associated Lymphoid Tissue-type)

[7, 8]. SS patients also have an 18.8 (confidence interval 9.5–37.3) times increased risk in developing lymphomas [9]. B cell lymphomas are most frequently located in the parotid gland. The conversion from variable to persistent enlargement of the gland is an alarming clinical sign. The presence of palpable purpura (fig. 5), vasculitis (fig. 6), renal involvement and peripheral neuropathy, although not pathognomonic, should raise suspicion, especially when it is combined with monoclonal gammopathy, reduced levels of complement C4, CD4+ T lymphocytopenia, a sharp increase in IgG levels or cryoglobulinemia [7–12] (table 2).

Serological Findings

The most characteristic autoantibodies in SS are the anti-Ro/SSA and anti-La/SSB autoantibodies (table 3), which are present in 70 and 50% of cases, respectively. Their titers reflect disease activity, while high titers of particularly anti-La/SSB have been associated with extraglandular disease [13]. Despite the fact that anti-Ro/SSA and anti-La/SSB are not specific for SS, since they can also occur in e.g. patients with systemic lupus erythematosus, their presence should alert the clinician for the possibility of a diagnosis of SS.

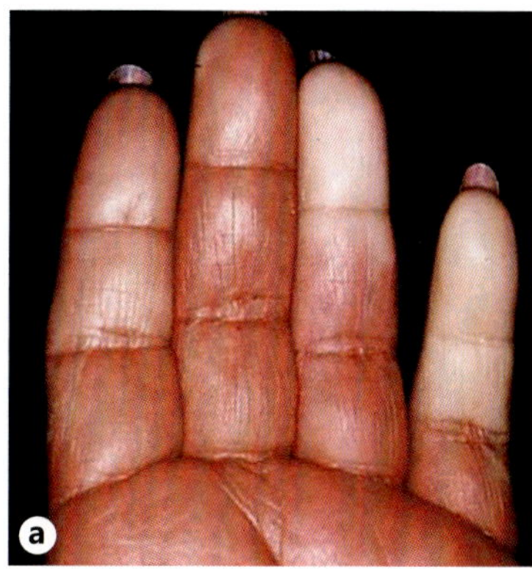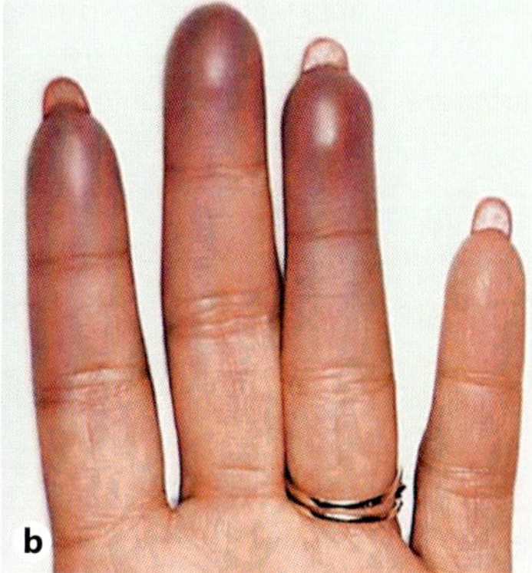

Fig. 6. Raynaud's phenomenon in two different cases (**a** and **b**).

Table 2. Risk factors for the development of lymphoma

Clinical	Serological
Persistent enlargement of parotid gland	Monoclonal gammopathy
Persistent lymphadenopathy	Reduced levels of complement C4
Palpable purpura	CD4+ T lymphocytopenia
Vasculitis	Increase in IgG levels
Renal involvement	Cryoglobulinemia
Peripheral neuropathy	

Table 3. Serological findings in SS patients

Autoantibody	Frequency in SS patients, %	Specific for SS
Anti-Ro/SSA	70	no
Anti-La/SSB	50	no
Anti-α-fodrin	30	yes
Antimuscarinic acetylcholine receptor 3	71–90	no
Rheumatoid factor	50	no

Diagnostic Criteria

In 2002, a joint study of the AECG presented the revised AECG classification criteria for SS and to date they are the most widely accepted criteria [2]. These criteria successfully combine subjective symptoms, as well as objective signs of keratoconjunctivitis sicca and hyposalivation together with histopathological and serological findings. It must be underlined that SS can be present in a patient who does not completely fulfil these classification criteria and that, since anticholinergic drugs are widely used by patients for many conditions, their exclusion should be carefully re-evaluated [1]. Recently, due to the emergence of biological agents, the American College of Rheumatology (ACR) proposed new classification criteria for SS, based merely on objective tests. The ACR classification criteria were developed from registry data collected with standardized measures and are thought to be more suitable in situations where misclassification may present health risks [14].

Etiopathogenesis

SS is considered to be an autoimmune disorder but with a pathogenesis that is poorly understood. It is not clarified yet whether this disturbance of the immune system exists primarily or is a result of an infection, possibly viral, or has another cause. Several findings suggest that viruses, such as Epstein-Barr virus, coxsackievirus and retroviruses may be implicated. Possibly, their glandular persistence in salivary gland epithelial cells may lead to chronic lymphocytic sialoadenitis with formation of foci around the ducts [15]. Additionally, hepatitis C and HIV infections can produce both symptoms and pathological findings similar to those in SS, but for the time being the presence of these infections is an exclusion criterion for SS.

Moreover various endogenous factors may be involved (table 4). The strong female preponderance implies possible involvement of hormonal factors, while the extended haplotype HLA-DR3/DQ2 in combination with the C4null gene being present in 50% of SS patients probably evinces genetic factors too [16, 17].

The complexity of the pathogenetic pathways in SS involves both systemic B cell hyperactivity and T cell lymphocytes targeting glandular epithelial cells:

- prolonged B cell survival and B cell hyperactivity lead to the presence of anti-Ro/SSA and anti-La/SSB antibodies, rheumatoid factor, type 2 cryoglobulins and hypergammaglobulinemia in SS patients;
- ductal epithelial cells are surrounded by activated T cells, predominantly CD4+ (70–80%); CD8+ T cells constitute around 10% of infiltrating cells in affected labial salivary glands [18].

Histopathology

Biopsy of the minor salivary glands of the lower lip is widely used for the diagnosis of SS, and its histopathology is considered as 1 of the 4 objective AECG classification criteria of SS as well as 1 of the 3 objective ACR criteria [2, 4]. The procedure is performed under local anesthesia. A lower lip mucosa incision of approximately 1.5 cm is made and at least 7 individual labial glands are collected [19]. Parotid biopsies are increasingly gaining broader acceptance and use as an alternative to minor salivary gland biopsies. Parotid biopsies are validated for the AECG classification criteria [20], but not yet for the ACR classification criteria. With this technique, parotid tissue is taken under local anesthesia, from the area around the lower ear lobe. A 1-cm incision is carried out, followed by blunt dissection to the parotid gland and an incisional biopsy. The remaining wound is closed in layers [21] (fig. 7). Several studies show a lower morbidity after a parotid gland biopsy compared to a lip biopsy [20, 22] with regard to loss of sensibility and pain. In none of the parotid gland biopsy studies was a disturbance of the facial nerve observed.

The most prominent microscopic finding in SS is periductal lymphocytic infiltration of salivary glands in combination with destruction of acini (fig. 8a). The infiltrates contain B and T lymphocytes as well as nonlymphoid cells and are located around the striated ducts. When these infiltrates are composed of more than 50 cells, they

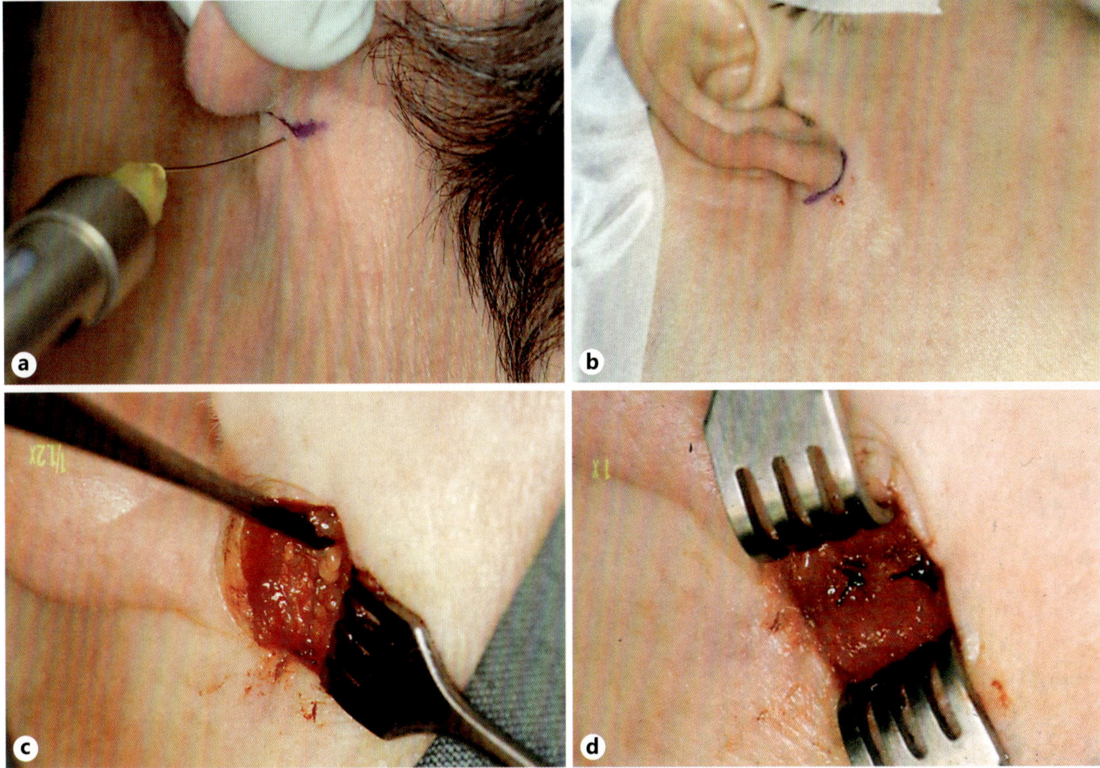

Fig. 7. Incisional biopsy of the parotid gland. **a** The area is anesthetized with local infiltration anesthesia. **b** With a No 15 blade a small 1–2 cm incision is made just below and behind the earlobe near the posterior angle of the mandible. **c** The skin is incised and the parotid capsule is exposed by blunt dissection. The capsule of the gland is carefully opened and a small amount of superficial parotid tissue is removed. **d** The procedure is completed with a 2 to 3-layered closure with 4-0 gauge absorbable sutures (polyglycolic acid), while the skin layer is closed with 5-0 nylon sutures.

are called focus. The presence of more than 1 focus per 4-mm^2 area of glandular tissue is regarded as a positive criterion for the diagnosis of SS. Furthermore, if the major glands are enlarged, progression to a lymphoepithelial lesion (fig. 8b) can also be present. In major salivary glands characteristic epimyoepithelial islands in a background of lymphoid stroma are usually seen.

The sensitivity and specificity of the parotid biopsy are comparable with those of labial salivary glands [20] and additionally can provide evidence about lymphoepithelial lesions and well-formed lymphoid follicles or germinal centers. It is suggested that the presence of germinal center-like structures in primary SS salivary biopsies is a highly predictive and easy-to-obtain marker for non-Hodgkin lymphoma development, allowing for risk stratification of patients and the possibility to initiate preventive B-cell-directed therapy [23]. The histopathological results of a parotid biopsy can be indicative of malignant lymphoma as MALT lymphomas often develop in the parotid gland and rarely in labial glands. Repeated biopsies from the same parotid gland offer information concerning the course of the disease.

Treatment

Evidence-based therapy for SS is limited, and the treatment of patients with SS is mainly supportive (table 5).

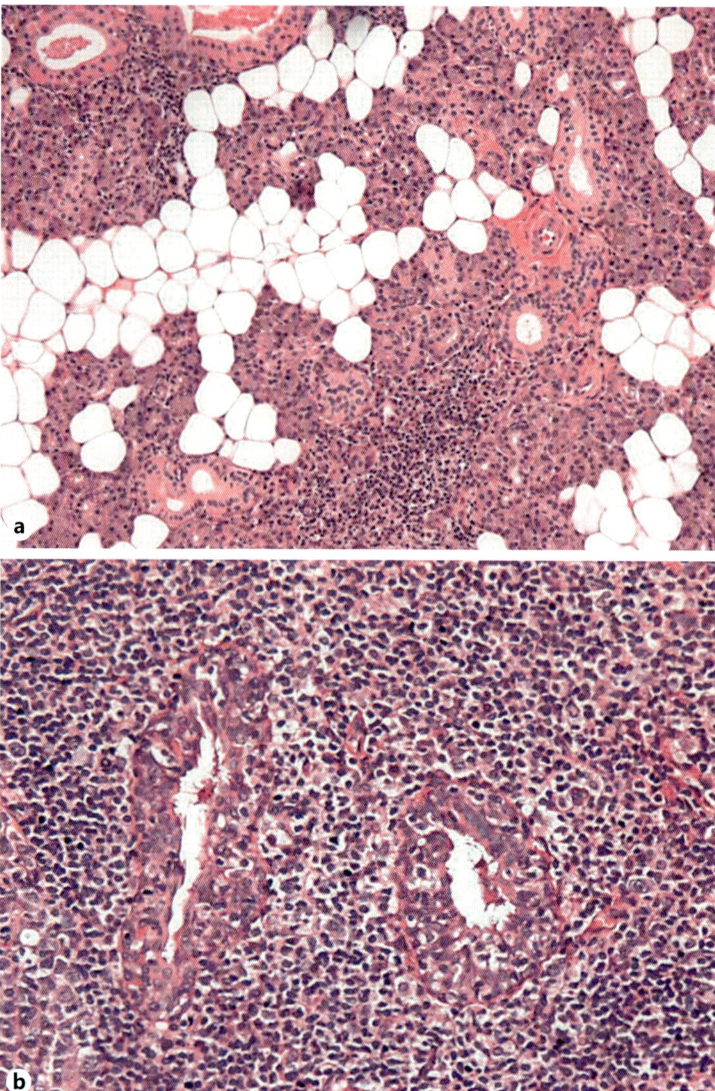

Fig. 8. a Parotid gland tissue showing a periductular lymphocytic infiltrate and deposition of fat between the serous acini. Magnification ×10. **b** Lymphoepithelial lesions surrounded by a lymphocytic infiltrate (from Pijpe et al. [20]). Magnification ×20.

Local treatment for dryness of eyes and mouth is helpful in many cases. Artificial tears lubricate dry eyes, and in case of keratoconjunctivitis local corticosteroids and local immunosuppressive agents may be used. Sealed glasses are also introduced in an attempt to prevent evaporation of tears and to conserve the tear film. Sealing the lacrimal punctum in the inner margin of the eyelid can also be helpful by blocking the normal drainage to the nose. To treat xerostomia, one first has to estimate whether stimulating salivary secretion by gustatory (sugar-free sweets), mechanical (chewing gum) or sialagogue medication (pilocarpine, cevimeline) results in relief of xerostomia. When stimulation of salivary secretion is uneventful, one can try to treat xerostomia with mouthrinses, artificial saliva and/or oral gels. Antifungal therapy, such as local treatment with nystatin, myconazole or amphotericin B, is frequently needed to treat oral candidia-

Supportive	Causal
(1) Local	(1) B cell depletion,
Eye	e.g. rituximab
Artificial tears	(2) Inhibition of costimulation of T cells,
Corticosteroids	e.g. abatacept
Immunosuppressives	(3) Anti-tumor necrosis factor α,
Sealed glasses	e.g. infliximab, etanercept
Blocking the lacrimal punctum	(4) Neutralization of interferon,
Mouth	e.g. rontalizumab
Artificial saliva	
Antifungal therapy	
Fluoride application	
(2) Systemic	
Pilocarpine	
Cevimeline	

sis. Due to the increased risk of dental caries, a weekly to daily use of topical neutral fluoride applications or mouthrinses is indicated in dentate patients.

During the past two decades biologicals have become available to target specific cells or cytokines that are fundamental in the immune response. Under this new perspective, inhibitors of tumor necrosis factor α, interferon α, B cell depletion therapies, B-cell-activating factor inhibitors and treatments targeting the costimulation of T cells have also been recruited in the treatment of SS [24, 25].

The therapeutic approach to the patients with SS and MALT lymphoma is a matter of debate [26].

Drugs

Drugs are the most common cause of dry mouth. A review of the 200 most frequently prescribed drugs in the USA revealed that the most common side effect was dry mouth (80.5%) followed by dysgeusia (47.5%) and stomatitis (33.9%) [27]. The exact relationship between dry mouth and drugs is variably influenced by many factors, such as type of drug, number of drugs, drug combination, dose, form, time of intake, duration of use, drug interaction and reliability of the patient's report. The situation is even more complicated in diseases and disorders that contribute to the problem. Nonetheless, it is generally accepted that the prevalence of dry mouth increases with age and the number of drugs taken per day.

Interrelation with Age and Sex
Studies indicate that the average intake of drugs increases with age. In the USA for example, the intake of 1–2 drugs/day/person progressively increases from 24 to 87% from the age of 18 to the age of 65, respectively (fig. 9). The prevalence of dry mouth increases with the number of drugs taken per day (fig. 10). Furthermore, drug-induced dryness is greater in women than men [28].

Possible Mechanisms
This section will focus on the mechanisms by which the most commonly used therapeutic drugs (table 6) induce xerostomia [29]. More detailed lists about xerogenic medication can be found at www.drymouth.info.

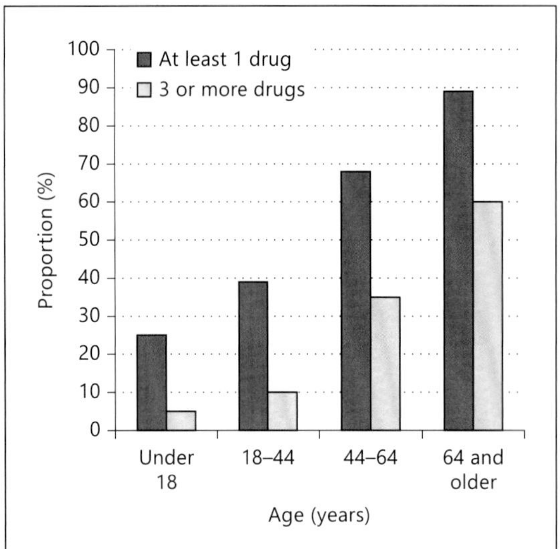

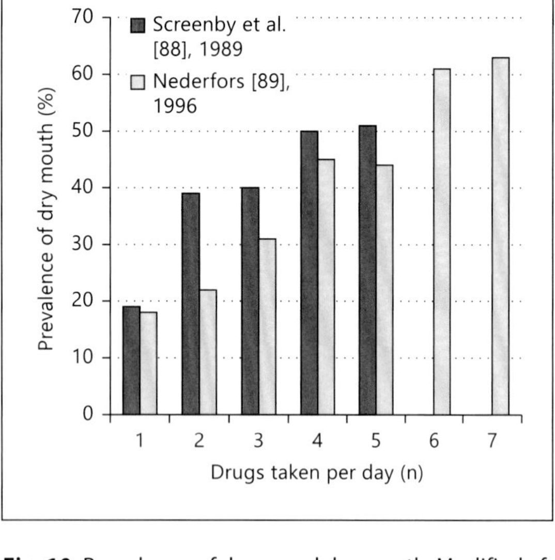

Fig. 9. Prescription drug use in the USA, 2001–2004. Modified after Sreenby and Vissink [1].

Fig. 10. Prevalence of drugs and dry mouth. Modified after Sreenby and Vissink [1].

Table 6. Top 10 therapeutic classes by US dispensed prescriptions in 2011 (http://www.imshealth.com)

Ranking	Drug class	Xerogenic effect
1	Antidepressants	yes
2	Lipid regulators	no
3	Narcotic analgesics	yes
4	Antidiabetics	no
5	Angiotensin-converting enzyme inhibitors	yes
6	β-Blockers	uncommon
7	Respiratory agents	yes
8	Antiulcerants	yes
9	Diuretics	yes
10	Antiepileptics	yes

Antidepressants

Antidepressants were the most common drug category prescribed in 2011 in the USA. Over 60% of patients prescribed antidepressants report taking these for more than 2 years, and 14% for 10 years or more. Antidepressants fall into 4 different classes, i.e. tricyclic antidepressants (TCAs), selective serotonin reuptake inhibitors, monoamine oxidase inhibitors and atypical antidepressants. The xerogenic capacity of TCAs is high. A meta-analysis of the adverse effects of TCAs versus selective serotonin reuptake inhibitors revealed that dry mouth occurred in 28% of patients taking TCAs in comparison to 7% of patients on selective serotonin reuptake inhibitors [30].

Narcotic Analgesics

Narcotic analgesics in general reduce neuronal excitability in the pain-carrying pathway by binding to opioid receptors. Narcotic analgesics used to alleviate acute and chronic pain are morphine and its analogues as well as some synthetic derivatives. Glare et al. [31] reported a prevalence of 95% of dry mouth among patients who received morphine due to cancer pain. The symptom was persistent with a moderate to severe intensity in 57% of them [31].

Angiotensin-Converting Enzyme Inhibitors

Angiotensin-converting enzyme inhibitors are widely used for the treatment of hypertension. They inhibit the conversion of angiotensin I to angiotensin II. Angiotensin II receptor antagonists block the binding of angiotensin II to the angiotensin receptors located in the vascular smooth muscle. Both angiotensin-converting enzyme inhibitors and angiotensin II receptor antagonists present similar actions and side effects. Reports suggest that about 1–13% of patients using these drugs complain of oral dryness [32, 33].

Respiratory Agents
Smidt et al. [34] reported that 27% of patients taking respiratory agents experience dry mouth. The main categories of respiratory agents associated with dry mouth are [34]:

drugs for obstructive airway diseases: the most commonly reported adverse event in patients on tiotropium (an antimuscarinic agent) was dry mouth (9.3 vs. 1.6%) relative to placebo [35], while overall 6–13% of patients receiving it complain of xerostomia [36, 37];

cough and cold preparations: stimulation of α- and β-adrenergic receptors in the mucous membrane of the respiratory tract by pseudoepinephrine and phenylephrine results in the shrinking of the nasal mucous membranes and relieves nasal obstruction; studies demonstrated that pseudoepinephrine induced oral dryness in 0.4–11% of patients [38, 39];

antihistamines: three antihistamines are commonly used, i.e. brompheniramine, clopheniramine and carbinoxamine; their frequent combination with decongestants as well as breathing through the mouth can cause dry mouth [38, 39].

Antiulcerants
Antiulcerants are used as part of the treatment of ulcers. The two basic types of antiulcerants are H_2 blockers and proton pump inhibitors. Dry mouth is found in 41% of patients receiving an H_2 receptor antagonist for eradication of *Helicobacter pylori* [40]. A subnormal parotid or whole-saliva flow rate is reported in patients treated with the proton pump inhibitor omeprazole [41].

Diuretics
Diuretics increase the formation and extraction of urine. As a result, there is a decrease in the volume of extracellular water and a consequent reduction in cardiac output and blood pressure. According to Smidt et al. [34], 17.8% of patients taking diuretics experience dry mouth. Habbab et al. [33] distinguished the frequency of dry mouth in patients taking thiazide, loop and potassium-sparing diuretics in 3, 8 and 16%, respectively.

Antiepileptics
Antiepileptics prevent rapid, repetitive stimulation of the brain that causes seizure activity. A selective dose-dependent pattern in the onset of dry mouth after the administration of pregabalin has been reported, which becomes evident after a dosage of 150 mg/day [42]. There are also sporadic reports of transient 'sicca syndrome' during phenobarbital treatment [43] or even phenytoin-induced pseudo-SS [44]. Mild xerogenic effects have been attributed to carbamazepine, oxcarbazepine, gabapentin, valproic acid, clonazepam, zonisamide, lamotrigine and topiramate (www. drymouth.info).

Management of Drug-Induced Xerostomia
Drug-induced xerostomia can be diminished by avoiding xerogenic drugs or minimizing their exposure to them. Substitution of a different agent with similar therapeutic properties can usually relieve the symptoms. If this is not possible, patients should be reassured that in most cases this condition is not permanent and salivary gland function will return to pretreatment levels after the end of the therapy. In order to support these patients, usage of salivary stimulants or artificial saliva, in particular substitutes with a stimulating additive such as malic acid (during daytime) and gel type substitutes (during the night), should be encouraged during their treatment with xerogenic drugs.

Radiotherapy

Radiotherapy plays a fundamental role in the treatment of the majority of patients with head and neck cancer. It can be used as a single modality or in combination with surgery and/or chemotherapy and typically involves administration of high doses to the major salivary glands. Ablation therapy of thyroid cancer with radioactive iodine treatment can also result in radiation damage to salivary gland tissue as, besides thyroid glands, salivary glands have a high uptake of this agent too.

In most cases radiation damage to salivary gland tissue results in progressive loss of glandular function and diminished salivary output. Patients complain of oral dryness, impairment of oral functions (speech, chewing and swallowing) because of insufficient lubrication of mucosal surfaces and of ingested food [45]. The oral mucosa can become dry and atrophic, leading to frequent ulceration and injury. The shift in oral microflora towards cariogenic bacteria in combination with the reduced saliva flow and altered saliva composition may lead to radiation caries [46, 47]. It must be noted that the subjective symptom of xerostomia does not always correlate with salivary flow rates; this not only applies to radiation-induced xerostomia, but also to xerostomia from other origins.

Pathophysiology
One week after the onset of conventional radiotherapy treatment, when 5–10 Gy are typically delivered, the salivary output declines by 60–90%. The acute phase of xerostomia is characterized by thick and sticky saliva, as a result of the faster decline in the serous, watery content of the saliva, compared to the decline of mucins and proteins. Late recovery is possible in cases of moderate radiation mode [48–50]. More recent studies revealed, however, that the serous parotid and seromucous submandibular gland are probably equally sensitive to ionizing radiation [51].

The mechanism of acute salivary damage is not fully understood, and to date several theories have been proposed, amongst others as described below.

(1) DNA damage caused by radiation impairs proper cell division, resulting in cell death or senescence of cells that attempt to divide. Since cells of salivary glands have a slow turnover rate (60–120 days), they are expected to be late responding tissue (>60 days) [52]. However, the changes in quantity and composition of saliva occur shortly after the radiation, indicating that salivary glands respond acutely [51, 53]. Radiation injury leads to the loss of saliva-producing acinar cells, but ducts, although deprived of function, remain intact [54]. The role of apoptotic cell death after radiotherapy is controversial. Paardenkooper et al. [55] did not observe a dose-related increase in apoptotic cells early after radiation therapy, whereas Avila et al. [56] found that early radiation-induced salivary gland dysfunction resulted from p53-dependent apoptosis. Currently, research focuses on very selectively blocking certain areas of the parotid gland from radiation injury meanwhile sparing those areas of the parotid gland where the stem cells reside (probably the main excretory ducts).

(2) The leakage of granules and subsequent lysis of acinar cells has been suggested as an alternative explanation for this phenomenon [57, 58]. Nevertheless, studies show no cell loss during the first days after irradiation [51, 59–61].

Management
Reducing the volume of irradiated salivary glands by advanced radiotherapy techniques in combination with salivary protectors and/or stimulators can be highly beneficial for patients.

Advanced Radiation Delivery Techniques
Prevalence rates of xerostomia after radiotherapy with conventional and more advanced techniques are shown in figure 11, of which 3-dimensional conformal radiotherapy (3-D-CRT) and intensity-modulated radiotherapy (IMRT) are currently most commonly applied.

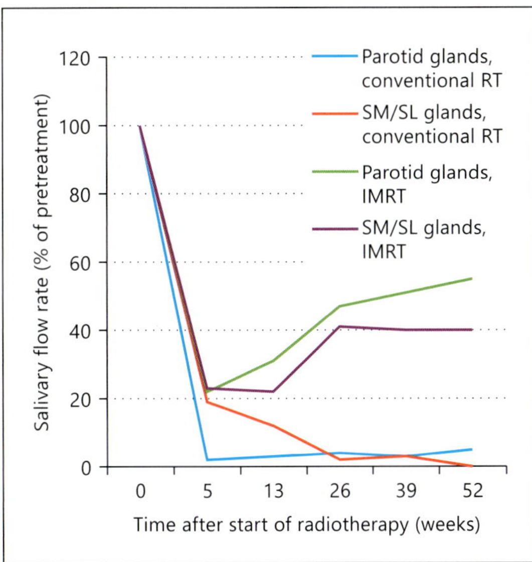

Fig. 11. Stimulated parotid and submandibular/sublingual (SM/SL) saliva flow rates after conventional radiotherapy (RT) and parotid-sparing intensity-modulated radiotherapy (IMRT). Modified after Vissink et al. [45].

3-D-CRT is designed to deliver an exact dose of irradiation to a target volume. This is achieved by creating a 3-dimensional image of the tumor so that multiple radiation beams can be shaped exactly to the contour of the treatment area. There is evidence that reduced radiotherapy dosages by 3-D-CRT to contralateral parotid glands result in less loss of salivary gland function after radiotherapy up to 2 years after the completion of radiotherapy [62]. Albeit 3-D-CRT has the potential to decrease the prevalence and severity of xerostomia, xerostomia has been shown to be significantly worse after bilateral compared to unilateral treatment.

IMRT is currently recommended as a standard approach in head and neck cancer, as it allows a more accurate distribution of specific radiation dosage and dosage distribution to the tumor and therefore provides better sparing of the surrounding tissues. Since it reduces the dose to salivary glands, it can contribute to the maintenance of adequate saliva flow rates and the reduction of xero-

stomia [63–65]. IMRT compared to 2-dimensional radiotherapy results in a significant decrease in xerostomia (both patient- and observer-rated). Approximately 40% of patients still complain of dry mouth [66].

A rather new technique that is yet sparsely applied in the clinic is *proton radiotherapy.* This technique uses charged particles (protons) instead of photons. The physical and radiobiological properties of protons allow a better dose distribution, compared with photon radiotherapy. Thus, the dose to normal tissues as well as the late side effects are minimized. The existing literature shows that the dose to critical organs can be significantly reduced, especially in patients with tumors located in the pharynx [67, 68] and the paranasal sinuses [69, 70] as well as in the head and neck cancer patients treated with bilateral neck irradiation [71].

Agents for Prevention of Xerostomia or Restoration of Lubrication
Pilocarpine
Pilocarpine is a cholinergic parasympathomimetic agent, acting as an agonist at muscarinic receptors. One third to two thirds of patients with postradiotherapy xerostomia can benefit from the administration of pilocarpine [72, 73]. A dose of 5 mg t.i.d. is recommended, and up to 4 weeks might be required before a maximum effect is visible. The possible mechanism involves stimulation of the residual function of the major salivary glands as well as the stimulation of the minor salivary glands, especially the ones in the palate, which have a greater resistance to irradiation [74].

The effects of postirradiation pilocarpine disappear when patients stop using it. In order to protect salivary glands during radiotherapy and to eliminate the long-term postirradiation effects, administration of pilocarpine during radiotherapy is an alternative choice [75, 76]. The beneficial effect of pilocarpine depends on the dose distribution in the parotid glands and when the parotid dose exceeds

40 Gy, administration of pilocarpine during radiotherapy can considerably spare parotid flow and reduce patient-rated xerostomia [77].

Amifostine

Direct radioprotection can be achieved by the use of amifostine, a scavenger of free radicals [78]. Although salivary flow is preserved when amifostine is concurrently delivered with radiation, patients continue to experience xerostomia. Intravenous administration is accompanied by several side effects (e.g. nausea, vomiting, hypotension). Furthermore amifostine might also have the undesirable effect of tumor protection. Thus, the debate continues whether it is safe to use it in cancer patients [79].

Tempol

Tempol is a stable nitroxide, providing radioprotection possibly by mimicking superoxide dismutase activity and scavenging free radicals. In a mouse model tempol significantly reduced salivary gland hypofunction [80], by protecting salivary glands, but not tumor tissue [81]. Thus, tempol could be considered for human clinical trials.

Keratinocyte Growth Factor

Keratinocyte growth factor can be administered prior to or during radiotherapy. It suppresses apoptosis and enhances survival and proliferation of salivary acinar cells [82]. Postirradiation administration of keratinocyte growth factor most likely accelerates expansion of the pool of progenitor/stem cells that survived the treatment.

Oral Lubricants and Saliva Substitutes

A symptomatic approach of xerostomia is attempted when stimulation of residual secretion is insufficient or in cases where there is a contraindication in the administration of the aforementioned agents. The most commonly applied and best-studied saliva substitutes are based on carboxymethylcellulose [83], mucin [84] or xanthan gum [85]. Mucin- and xanthan gum-containing substitutes are usually preferred because they have superior rheological and wetting properties compared to carboxymethylcellulose-based saliva substitutes. During the night and when daily activities are at a low level, gel-like saliva substitutes are preferred [86].

Adult Salivary Gland Stem Cells

Supportive and palliative care does not address the source of the problem: a lack of functional saliva-producing cells, caused by radiation-induced stem cell sterilization. Recent identification of salivary gland stem and progenitor cell populations provides a basis for the development of a stem cell-based therapy for xerostomia [87].

References

1 Sreenby LM, Vissink A: Dry Mouth. The Malevolent Symptom: A Clinical Guide, ed 1. Iowa, Wiley-Blackwell, 2010.

2 Vitali C, Bombardieri S, Jonsson R, Moutsopoulos HM, Alexander EL, Carsons SE, Daniels TE, Fox PC, Fox RI, Kassan SS, Pillemer SR, Talal N, Weisman MH, European Study Group on Classification Criteria for Sjögren's Syndrome: Classification criteria for Sjögren's syndrome: a revised version of the European criteria proposed by the American-European Consensus Group. Ann Rheum Dis 2002; 61:554–558.

3 Kassan SS, Moutsopoulos HM: Clinical manifestations and early diagnosis of Sjögren syndrome. More author information. Arch Intern Med 2004;164: 1275–1284.

4 Anaya JM, Talal N: Sjögren's syndrome; in Kassirer JP, Greene HL (eds): Current Therapy in Adult Medicine, ed 4. Baltimore, Mosby, 1997, pp 1291–1298.

5 Papiris SA, Maniati M, Constantopoulos SH, Roussos C, Moutsopoulos HM, Skopouli FN: Lung involvement in primary Sjögren's syndrome is mainly related to the small airway disease. Ann Rheum Dis 1995;58:61–64.

6 Al-Hashimi I, Khuder S, Haghighat N, Zipp M: Frequency and predictive value of the clinical manifestations in Sjögren's syndrome. J Oral Pathol Med 2001;30:1–6.

7 Theander E, Henriksson G, Ljungberg O, Mandl T, Manthorpe R, Jacobsson LT: Lymphoma and other malignancies in primary Sjögren's syndrome: a cohort study on cancer incidence and lymphoma predictors. Ann Rheum Dis 2006;65: 796–803.

8 Sutcliffe N, Inanc M, Speight P, Isenberg D: Predictors of lymphoma development in primary Sjögren's syndrome. Semin Arthritis Rheum 1998;28:80–87.

9 Zintzaras E, Voulgarelis M, Moutsopoulos HM: The risk of lymphoma development in autoimmune diseases: a meta-analysis. Arch Intern Med 2005;165: 2337–2344.

10 Ioannidis JP, Vassiliou VA, Moutsopoulos HM: Long-term risk of mortality and lymphoproliferative disease and predictive classification of primary Sjögren's syndrome. Arthritis Rheum 2002;46: 741–747.

11 Theander E, Manthorpe R, Jacobsson LT: Mortality and causes of death in primary Sjögren's syndrome: a prospective cohort study. Arthritis Rheum 2004; 50:1262–1269.

12 Fox RI: Sjögren's syndrome. Lancet 2005;366:321–331.

13 Maran R, Dueymes M, Pennec YL, Casburn-Budd R, Shoenfeld Y, Youinou P: Predominance of IgG1 subclass of anti-Ro/SSA, but not anti-La/SSB antibodies in primary Sjögren's syndrome. J Autoimmun 1993;6:379–387.

14 Shiboski SC, Shiboski CH, Criswell L, Baer A, Challacombe S, Lanfranchi H, Schiødt M, Umehara H, Vivino F, Zhao Y, Dong Y, Greenspan D, Heidenreich AM, Helin P, Kirkham B, Kitagawa K, Larkin G, Li M, Lietman T, Lindegaard J, McNamara N, Sack K, Shirlaw P, Sugai S, Vollenweider C, Whitcher J, Wu A, Zhang S, Zhang W, Greenspan J, Daniels T; Sjögren's International Collaborative Clinical Alliance (SICCA) Research Groups: American College of Rheumatology classification criteria for Sjögren's syndrome: a data-driven, expert consensus approach in the Sjögren's International Collaborative Clinical Alliance cohort. Arthritis Care Res (Hoboken) 2012;64:475–487.

15 Triantafyllopoulou A, Tapinos N, Moutsopoulos HM: Evidence for coxsackievirus infection in primary Sjögren's syndrome. Arthritis Rheum 2004;50:2897–2902.

16 Price EJ, Venables PJ: The etiopathogenesis of Sjögren's syndrome. Semin Arthritis Rheum 1995;25:117–133.

17 Hansen A, Lipsky PE, Dörner T: Immunopathogenesis of primary Sjögren's syndrome: implications for disease management and therapy. Curr Opin Rheumatol 2005;17:558–565.

18 Tapinos NI, Polihronis M, Tzioufas AG, Skopouli FN: Immunopathology of Sjögren's syndrome. Ann Méd Interne (Paris) 1998;149:17–24.

19 Greenspan JS, Daniels TE, Talal N, Sylvester RA: The histopathology of Sjögren's syndrome in labial salivary gland biopsies. Oral Surg Oral Med Oral Pathol 1974;37:217–229.

20 Pijpe J, Kalk WW, van der Wal JE, Vissink A, Kluin PM, Roodenburg JL, Bootsma H, Kallenberg CG, Spijkervet FK: Parotid gland biopsy compared with labial biopsy in the diagnosis of patients with primary Sjögren's syndrome. Rheumatology (Oxford) 2007;46:335–341.

21 Kraaijenhagen HA: Letter: technique for parotid biopsy. J Oral Surg 1975;33:328.

22 Marx RE, Hartman KS, Rethman KV: A prospective study comparing incisional labial to incisional parotid biopsies in the detection and confirmation of sarcoidosis, Sjögren's disease, sialosis and lymphoma. J Rheumatol 1988;15:621–629.

23 Theander E, Vasaitis L, Baecklund E, Nordmark G, Warfvinge G, Liedholm R, Brokstad K, Jonsson R, Jonsson MV: Lymphoid organisation in labial salivary gland biopsies is a possible predictor for the development of malignant lymphoma in primary Sjögren's syndrome. Ann Rheum Dis 2011;70:1363–1368.

24 Cummins MJ, Papas A, Kammer GM, Fox PC: Treatment of primary Sjögren's syndrome with low-dose human interferon alfa administered by the oromucosal route: combined phase III results. Arthritis Rheum 2003;49:585–593.

25 Meiners PM, Vissink A, Kallenberg CG, Kroese FG, Bootsma H: Treatment of primary Sjögren's syndrome with anti-CD20 therapy (rituximab). A feasible approach or just a starting point? Expert Opin Biol Ther 2011;11:1381–1394.

26 Pollard RP, Pijpe J, Bootsma H, Spijkervet FK, Kluin PM, Roodenburg JL, Kallenberg CG, Vissink A, van Imhoff GW: Treatment of mucosa-associated lymphoid tissue lymphoma in Sjögren's syndrome: a retrospective clinical study. J Rheumatol 2011;38:2198–2208.

27 Smith RG, Burtner AP: Oral side-effects of the most frequently prescribed drugs. Spec Care Dentist 1994;14:96–102.

28 Nederfors T, Isaksson R, Mörnstad H, Dahlöf C: Prevalence of perceived symptoms of dry mouth in an adult Swedish population – relation to age, sex and pharmacotherapy. Community Dent Oral Epidemiol 1997;25:211–216.

29 Scully C: Drug effects on salivary glands: dry mouth. Oral Dis 2003;9:165–176.

30 Wilson K, Mottram P: A comparison of side effects of selective serotonin reuptake inhibitors and tricyclic antidepressants in older depressed patients: a meta-analysis. Int J Geriatr Psychiatry 2004;19:754–762.

31 Glare P, Walsh D, Sheehan D: The adverse effects of morphine: a prospective survey of common symptoms during repeated dosing for chronic cancer pain. Am J Hosp Palliat Care 2006;23:229–235.

32 Mangrella M, Motola G, Russo F, Mazzeo F, Giassa T, Falcone G, Rossi F, D'Alessio O, Rossi F: Hospital intensive monitoring of adverse reactions of ACE inhibitors. Minerva Med 1998;89:91–97.

33 Habbab KM, Moles DR, Porter SR: Potential oral manifestations of cardiovascular drugs. Oral Dis 2010;16:769–773.

34 Smidt D, Torpet LA, Nauntofte B, Heegaard KM, Pedersen AM: Associations between oral and ocular dryness, labial and whole salivary flow rates, systemic diseases and medications in a sample of older people. Community Dent Oral Epidemiol 2011;39:276–288.

35 Casaburi R, Briggs DD Jr, Donohue JF, Serby CW, Menjoge SS, Witek TJ Jr: The spirometric efficacy of once-daily dosing with tiotropium in stable COPD: a 13-week multicenter trial. The US Tiotropium Study Group. Chest 2000;118: 1294–1302.

36 Durham MC: Tiotropium (Spiriva): a once-daily inhaled anticholinergic medication for chronic obstructive pulmonary disease. Proc (Bayl Univ Med Cent) 2004;17:366–373.

37 Keam SJ, Keating GM: Tiotropium bromide. A review of its use as maintenance therapy in patients with COPD. Treat Respir Med 2004;3:247–268.

38 Kaiser HB, Banov CH, Berkowitz RR, Bernstein DI, Bronsky EA, Georgitis JW, Mendelson LM, Rooklin AR, Sholler LJ, Stricker WW, Harrison JE, Danzig MR, Lorber RR: Comparative efficacy and safety of once-daily versus twice-daily loratadine-pseudoephedrine combinations versus placebo in seasonal allergic rhinitis. Am J Ther 1998;5:245–251.

39 Wellington K, Jarvis B: Cetirizine/pseu-doephedrine. Drugs 2001;61:2231–2242.
40 Kaviani MJ, Malekzadeh R, Vahedi H, Sotoudeh M, Kamalian N, Amini M, Massarrat S: Various durations of a standard regimen (amoxycillin, metro-nidazole, colloidal bismuth sub-citrate for 2 weeks or with additional ranitidine for 1 or 2 weeks) on eradication of *Helicobacter pylori* in Iranian peptic ulcer patients. A randomized controlled trial. Eur J Gastroenterol Hepatol 2001;13: 915–919.
41 Teare JP, Spedding C, Whitehead MW, Greenfield SM, Challacombe SJ, Thompson RP: Omeprazole and dry mouth. Scand J Gastroenterol 1995;30:216–218.
42 Zaccara G, Gangemi P, Perucca P, Specchio L: The adverse event profile of pregabalin: a systematic review and meta-analysis of randomized controlled trials. Epilepsia 2011;52:826–836.
43 Marino D, Malandrini A, Rocchi R, Selvi E, Federico A: Transient 'sicca syndrome' during phenobarbital treatment. J Neurol Sci 2011;300:164.
44 Chakravarty K, al-Jafari MS: Phenytoin-induced pseudo-Sjögren's syndrome. Br J Rheumatol 1996;35:1033–1034.
45 Vissink A, Mitchell JB, Baum BJ, Limesand KH, Jensen SB, Fox PC, Elting LS, Langendijk JA, Coppes RP, Reyland ME: Clinical management of salivary gland hypofunction and xerostomia in head-and-neck cancer patients: successes and barriers. Int J Radiat Oncol Biol Phys 2010;78:983–989.
46 Jensen SB, Pedersen AM, Reibel J, Nauntofte B: Xerostomia and hypofunction of the salivary glands in cancer therapy. Support Care Cancer 2003;11: 207–225.
47 Vissink A, Jansma J, Spijkervet FK, Burlage FR, Coppes RP: Oral sequelae of head and neck radiotherapy. Crit Rev Oral Biol Med 2003;14:199–212.
48 Shannon IL, Trodahl JN, Starcke EN: Radiosensitivity of the human parotid gland. Proc Soc Exp Biol Med 1978;157: 50–53.
49 Mossman KL: Quantitative radiation dose-response relationships for normal tissues in man. II. Response of the salivary glands during radiotherapy. Radiat Res 1983;95:392–398.
50 Ship JA, Eisbruch A, D'Hondt E, Jones RE: Parotid sparing study in head and neck cancer patients receiving bilateral radiation therapy: one-year results. J Dent Res 1997;76:807–813.

51 Coppes RP, Zeilstra LJW, Kampinga HH, Konings AW: Early to late sparing of radiation damage to the parotid gland by adrenergic and muscarinic receptor agonists. Br J Cancer 2001;85:1055–1063.
52 Stewart FA, Van der Kogel AJ: Volume effects in normal tissues; in Steel GG (ed): Basic Clinical Radiology, ed 3. London, Arnold, 2002, pp 42–51.
53 Burlage FR, Coppes RP, Meertens H, Stokman MA, Vissink A: Parotid and submandibular/sublingual salivary flow during high dose radiotherapy. Radiother Oncol 2001;61:271–274.
54 Konings AW, Coppes RP, Vissink A: On the mechanism of salivary gland radio-sensitivity. Int J Radiat Oncol Biol Phys 2005;62:1187–1194.
55 Paardenkooper GMRM, Zeilstra LJW, Coppes RP, Coppes RP, Konings AW: Radiation induced apoptosis is not the cause of acute impairment of rat salivary gland function. Int J Radiat Biol 1998; 73:641–648.
56 Avila JL, Grundmann O, Burd R, Limesand KH: Radiation-induced salivary gland dysfunction results from p53-dependent apoptosis. Int J Radiat Oncol Biol Phys 2009;73:523–529.
57 Abok K, Brunk U, Jung B, Ericsson J: Morphologic and histochemical studies of the deferring radiosensitivity of the ductular and acinar cell of the rat submandibular gland. Virchows Arch Cell Pathol 1984;45:443–460.
58 Nagler RM, Marmary Y, Fox PC, Baum BJ, Har-El R, Chevion M: Irradiation-induced damage to the salivary glands: the role of redox-active iron and copper. Radiat Res 1997;147:468–475.
59 Vissink A, 's-Gravenmade EJ, Ligeon EE, Konings WT: A functional and chemical study of radiation effects on rat parotid and submandibular/sublingual glands. Radiat Res 1990;124:259–265.
60 Vissink A, Kalicharan D, 's-Gravenmade EJ, Jongebloed WL, Ligeon EE, Nieuwenhuis P, Konings AW: Acute irradiation effects on morphology and function of rat submandibular glands. J Oral Pathol Med 1991;20:449–456.
61 Zeilstra LJ, Vissink A, Konings AWT, Coppes RP: Radiation induced cell loss in rat submandibular gland and its relation to gland function. Int J Radiat Biol 2000;76:419–429.

62 Jensen SB, Pedersen AM, Vissink A, Andersen E, Brown CG, Davies AN, Dutilh J, Fulton JS, Jankovic L, Lopes NN, Mello AL, Muniz LV, Murdoch-Kinch CA, Nair RG, Napeñas JJ, Nogueira-Rodrigues A, Saunders D, Stirling B, von Bültzingslöwen I, Weikel DS, Elting LS, Spijkervet FK, Brennan MT; Salivary Gland Hypofunction/Xerostomia Section, Oral Care Study Group, Multinational Association of Supportive Care in Cancer (MASCC)/International Society of Oral Oncology (ISOO): A systematic review of salivary gland hypofunction and xerostomia induced by cancer therapies: prevalence, severity and impact on quality of life. Support Care Cancer 2010;18:1039–1060.
63 Eisbruch A, Ten Haken RK, Hyungjin MK, Marsh LH, Ship JA: Dose, volume and function relationships in parotid salivary glands following conformal and intensity-modulated irradiation of head and neck cancer. Int J Radiat Oncol Biol Phys 1999;45:577–587.
64 Roesink JM, Moerland MA, Battermann JJ, Hordijk GJ, Terhaard CH: Quantitative dose-volume response analysis of changes in parotid gland function after radiotherapy in the head-and-neck region. Int J Radiat Oncol Biol Phys 2001; 51:938–946.
65 Murdoch-Kinch CA, Kim HM, Vineberg KA, Ship JA, Eisbruch A: Dose-effect relationship for the submandibular salivary glands and implications for their sparing by intensity modulated radiotherapy. Int J Radiat Oncol Biol Phys 2008;72:373–382.
66 Vergeer MR, Doornaert PA, Rietveld DH, Leemans CR, Slotman BJ, Langendijk JA: Intensity- modulated radiotherapy reduces radiation-induced morbidity and improves health-related quality of life: results of a nonrandomized prospective study using a standardized follow-up program. Int J Radiat Oncol Biol Phys 2009;74:1–8.
67 Steneker M, Lomax A, Schneider U: Intensity modulated photon and proton therapy for the treatment of head and neck tumors. Radiother Oncol 2006;80: 263–267.
68 Widesott L, Pierelli A, Fiorino C, et al: Intensity-modulated proton therapy versus helical tomotherapy in nasopharynx cancer: planning comparison and NTCP evaluation. Int J Radiat Oncol Biol Phys 2008;72:589–596.

   Delli · Spijkervet · Kroese · Bootsma · Vissink

69 Cozzi L, Fogliata A, Lomax A, Bolsi A: A treatment planning comparison of 3D conformal therapy, intensity modulated photon therapy and proton therapy for treatment of advanced head and neck tumours. Radiother Oncol 2001;61:287–297.

70 Lomax AJ, Goitein M, Adams J: Intensity modulation in radiotherapy: photons versus protons in the paranasal sinus. Radiother Oncol 2003;66:11–18.

71 Van de Water T, Lomax T, Bijl HP, Schilstra K, Hug E, Langendijk JA: Comparative treatment planning study between scanned intensity modulated proton therapy and photon therapy in complex oropharyngeal carcinoma. Radiother Oncol 2008;88:S77–S78.

72 Johnson JT, Ferretti GA, Nethery WJ, Valdez IH, Fox PC, Ng D, Muscoplat CC, Gallagher SC: Oral pilocarpine for postirradiation xerostomia in patients with head and neck cancer. N Engl J Med 1993;329:390–395.

73 LeVeque FG, Montgomery M, Potter D, Zimmer MB, Rieke JW, Steiger BW, Gallagher SC, Muscoplat CC: A multicenter, randomized, double-blind, placebo-controlled, dose-titration study of oral pilocarpine for treatment of radiation-induced xerostomia in head and neck cancer patients. J Clin Oncol 1993;11:1124–1131.

74 Niedermeier W, Matthaeus C, Meyer C, Staar S, Müller RP, Schulze HJ: Radiation-induced hyposalivation and its treatment with oral pilocarpine. Oral Surg Oral Med Oral Pathol Oral Radiol Endod 1998;86:541–549.

75 Warde P, O'Sullivan B, Aslanidis J, Kroll B, Lockwood G, Waldron J, Payne D, Bayley A, Ringash J, Kim J, Liu FF, Maxymiw W, Sprague S, Cummings BJ: A phase III placebo-controlled trial of oral pilocarpine in patients undergoing radiotherapy for head-and-neck cancer. Int J Radiat Oncol Biol Phys 2002;54:9–13.

76 Scarantino C, LeVeque F, Swann RS, White R, Schulsinger A, Hodson DI, Meredith R, Foote R, Brachman D, Lee N: Effect of pilocarpine during radiation therapy: results of RTOG 97-09, a phase III randomized study in head and neck cancer patients. J Support Oncol 2006;4:252–258.

77 Burlage FR, Roesink JM, Kampinga HH, Coppes RP, Terhaard C, Langendijk JA, van Luijk P, Stokman MA, Vissink A: Protection of salivary function by concomitant pilocarpine during radiotherapy: a double-blind, randomized, placebo-controlled study. Int J Radiat Oncol Biol Phys 2008;70:14–22.

78 Wasserman TH, Brizel DM, Henke M, Monnier A, Eschwege F, Sauer R, Strnad V: Influence of intravenous amifostine on xerostomia, tumor control, and survival after radiotherapy for head-and-neck cancer: 2 year follow-up of a prospective, randomized, phase III trial. Int J Radiat Oncol Biol Phys 2005;63:985–990.

79 Brizel DM, Wasserman TH, Henke M, Strnad V, Rudat V, Monnier A, Eschwege F, Zhang J, Russell L, Oster W, Sauer R: Phase III randomized trial of amifostine as a radioprotector in head and neck cancer. J Clin Oncol 2000;18:3339–3345.

80 Cotrim AP, Hyodo F, Matsumoto K, Sowers AL, Cook JA, Baum BJ, Krishna MC, Mitchell JB: Differential radiation protection of salivary glands versus tumor by tempol with accompanying tissue assessment of tempol by magnetic resonance imaging. Clin Cancer Res 2007;13:4928–4933.

81 Cotrim AP, Sowers AL, Lodde BM, Vitolo JM, Kingman A, Russo A, Mitchell JB, Baum BJ: Kinetics of tempol for prevention of xerostomia following head and neck irradiation in a mouse model. Clin Cancer Res 2005;11:7564–7568.

82 Dreyer JO, Sakuma Y, Seifert G: Die Strahlen-Sialadenitis. Stadieneinteilung und Immunhistologie. Pathologe 1989;10:165–170.

83 Matzker J, Schreiber J: Synthetischer Speichel zur Therapie der Hyposalien insbesondere bei der radiogenen Sialadenitis. Z Laryngol Rinol Otol 1972;51:422–428.

84 's-Gravenmade EJ, Roukema DA, Panders AK: The effects of mucin-containing artificial saliva on severe xerostomia. Int J Oral Surg 1974;3:435–439.

85 Van der Reijden WA, Van der Kwaak H, Vissink A, Veerman ECI, Nieuw Amerongen AV: Treatment of xerostomia with polymer-based saliva substitutes in patients with Sjögren's syndrome. Arthritis Rheum 1996;39:57–69.

86 Regelink G, Vissink A, Reintsema H, Nauta JM: Efficacy of a synthetic polymer saliva substitute in reducing oral complaints of patients suffering from irradiation-induced xerostomia. Quintessence Int 1998;29:383–388.

87 Pringle S, Van Os R, Coppes RP: Concise review: adult salivary gland stem cells and a potential therapy for xerostomia. Stem Cells 2012;31:613–619.

Konstantina Delli
Department of Oral and Maxillofacial Surgery
PO Box 30001
NL–9700 RB Groningen (The Netherlands)
E-Mail k.delli@umcg.nl

Ligtenberg AJM, Veerman ECI (eds): Saliva: Secretion and Functions.
Monogr Oral Sci. Basel, Karger, 2014, vol 24, pp 126–134 (DOI: 10.1159/000358793)

Drooling

Francisco Javier Silvestre-Donat · Javier Silvestre-Rangil

Department of Stomatology, University of Valencia, Valencia, Spain

Abstract

The uncontrolled and continuous release of saliva from the mouth is known as drooling. While accepted as normal in young children up to 2 years of age, drooling in older children and adolescents is secondary to altered orofacial neuromuscular control during development, and in the elderly it is a consequence of neurodegenerative disease. The underlying cause is patient inability to seal the lips, excess salivation and the inability to adequately swallow saliva. The estimated mean prevalence of drooling in such elderly patients is 37%. Drooling can give rise to irritation and excoriation of the skin around the mouth or chin, favors infections and gives rise to speech or eating disorders. Observational methods based on collection of the leaked saliva or documentation of the affected skin zones can be used to measure drooling. The management of such patients requires a multidisciplinary approach and comprises myofunctional therapy, behavioral change techniques, the administration of antisialagogues, botulinum toxin, or the use of certain surgical techniques designed to reduce salivary secretion or to deviate it towards posterior areas of the oral cavity.

Drooling is the uncontrolled and continuous release of saliva from the mouth in children or adults, and may be due to a range of causes. It is accepted to be normal in small children during tooth eruption and until 18–24 months of age, but is considered abnormal beyond 4 years of age.

Children may experience drooling in relation to certain diseases accompanied by fever and swallowing difficulties, such as upper airway infections, peritonsillar or retropharyngeal abscesses, and tonsillitis. Sudden drooling can also develop as a result of pesticide poisoning, during epileptic episodes and as a reaction to the venom of certain insects or reptiles [1, 2].

In elderly people, drooling appears associated with neurodegenerative diseases, though it is more often seen in young patients with altered orofacial neuromuscular control during development.

Drooling has been regarded as synonymous of sialorrhea by many authors. However, strictly speaking sialorrhea is the sensation of abundant saliva, in the same way as xerostomia refers to subjective dry mouth sensation. In many cases we use the term sialorrhea in reference to salivary hypersecretion or hypersialia, and although excessive salivation may be a cause of drooling, the latter must be accompanied by patient inability to retain saliva in the mouth or inability to swallow it correctly – as a result of which saliva leaks from the mouth. For this reason, drooling has also been referred to as 'false sialorrhea' [3].

Table 1. The swallowing process

First phase (oral)
    Salivation prior to food intake
Second phase (oral)
    Food bolus formation, which is displaced to the pharynx by the tongue
    + Full lip sealing
    + Sensorial stimulation upon lip mucosa to start neuromuscular activity
Third phase (pharyngeal)
    Bolus is displaced through the upper esophageal sphincter
    + Hypopharynx triggers coordinated actions for velopharyngeal sealing
    + Coordinated contraction of the pharyngeal muscles
    + Hyoid bone and larynx elevation and leveling of epiglottis
Last phase (esophageal)
    Relaxation of the lower esophageal sphincter
    + Peristaltic waves to advance the food bolus to the stomach

Salivation and Swallowing Mechanisms

Saliva is mainly secreted by the major salivary glands and to a lesser degree by the minor salivary glands. The former consist of three pairs of symmetrical glands on either side of the face: the parotid glands produce watery serous saliva, fundamentally in response to stimulation during meals, while the submaxillary and sublingual glands – located on both sides of the mandible – continuously produce a more viscous and mucinous saliva throughout the day. The saliva leaking from the mouth in drooling patients is predominantly of this mucinous type [4].

Salivary secretion is regulated by a nonconditioned reflex. Following initial stimulation of different chemoreceptors and mechanoreceptors in the oral cavity, the afferent nerve signals reach the salivary glands in the central nervous system. In turn, the efferent action potentials reach the salivary glands through sympathetic and parasympathetic pathways, causing the secretion of saliva.

Swallowing in turn is a complex muscle phenomenon that transports the food bolus from the oral cavity to the stomach [5] (table 1). This function is regulated by different cranial nerves and can be triggered consciously, with the participation of about 30 muscles and 6 cranial nerves

(pairs V, VII, IX, XI, XII and the accessory nerve). Swallowing function manifests itself around week 8 of intrauterine development and is crucial for survival of the individual.

Swallowing comprises a series of rapid, coordinated movements involving a number of phases – beginning at the oral level in the form of salivation prior to food intake. This is followed by a voluntary phase in which the food bolus is formed, and after the required stimuli, the bolus is displaced towards the pharynx by the tongue. Full lip sealing is required in order to ensure that the food bolus advances properly, and this is one of the mechanisms that fail in drooling. In turn, sensory stimulation upon the lip mucosa favors the start of the neuromuscular activity.

A third phase in swallowing involves the pharynx, where the food bolus is propelled until it passes through the upper esophageal sphincter. Bolus-mediated stimulation of the hypopharynx triggers a series of coordinated actions such as velopharyngeal sealing to avoid food penetration into the nasal passages, relaxation of the esophageal sphincter, contraction of the pharyngeal constrictor muscles, forward elevation of the hyoid bone and larynx, and leveling of the epiglottis. The epiglottis in turn seals itself to prevent bolus aspiration towards the airway. These mechanisms

Table 2. The etiology of drooling

Neurological deficits	Orofacial problems
Cerebral palsy	Positioning of the head
Mental retardation	Breathing through the mouth
Rare syndromes	Open mouth
Parkinson's disease	Lip incompetency
Encephalitis	Tongue deformities
Hydrocephalus	Reabsorbed dental ridge
Congenital suprabulbar palsy	Malocclusion
Cerebrovascular events	Surgical defects
Facial paralysis	
Amyotrophic lateral sclerosis	
Drug reaction	
Sjögren's syndrome (early sign)	

require feedback in the form of sensory stimuli from the hypopharyngeal mucosa.

Lastly, the esophageal phase of swallowing is characterized by relaxation of the lower esophageal sphincter and the generation of a sustained esophageal peristaltic wave that advances and propels the food bolus towards the stomach.

Prevalence of Drooling

The prevalence of drooling fundamentally has been studied in patients with brain palsy, since it is particularly common in this population group. The estimated prevalence in brain palsy patients is between 16.8 and 58%. However, when considering only severe drooling, the reported prevalence is 15–33%. Likewise, drooling has been reported to decrease with age. In effect, in small children (brain palsy) with primary dentition the prevalence of drooling is reportedly 75%, while in individuals with permanent dentition the figure drops to 43% [6]. This variation is considered to be related to maturation of the orofacial neuromuscular system as the children grow older [6–9].

In patients with neurodegenerative changes such as Parkinson's disease, the prevalence of daytime drooling has been estimated to vary between 20.8 and 28%. In turn, a relationship has been described between drooling and certain oral functions such as speech, swallowing and eating difficulties. In this context, intense drooling is associated with severer dysarthria among the affected individuals.

Causes and Physiopathology of Drooling

Drooling can be observed in many clinical situations, though in general it is seen in patients with neurological alterations and/or local orofacial problems (table 2).

The neurological causes of drooling include cerebral palsy, severe mental retardation or rare syndromes such as Moebius' syndrome, Angelman's syndrome, Worster-Drought syndrome, Freeman-Sheldon syndrome, Riley-Day syndrome or Landau-Kleffner syndrome. Other neurological processes in which drooling can be observed include Parkinson's disease, encephalitis, hydrocephalus, congenital suprabulbar palsy, cerebrovascular events, facial paralysis and amyotrophic lateral sclerosis [1].

Drooling can also manifest as an adverse reaction to certain drugs (e.g. clozapine) or as an early sign of Sjögren's syndrome [1] and in some patients with psychosis. Drooling can likewise be observed in certain emotional states and has also been related to gastroesophageal reflux.

On the other hand, the orofacial causes of drooling comprise a very diverse range of situations. Correct positioning of the head can be decisive for preventing the leakage of saliva from the mouth. Likewise, situations characterized by obstruction of the nasal passages cause the patient to breathe through the mouth, and this in turn favors drooling. In general, all alterations that cause opening of the mouth facilitate drooling. On the other hand, pharyngeal or esophageal obstruction secondary to neoplastic disease or surgical defects after head and neck surgery can also cause drooling. Deficient lip sealing is one of the most important factors, however. Missing lower anterior teeth or atrophy or reabsorption of the alveolar processes at this level also favor drooling in the same way as certain malocclusions, tongue deformities and anesthesia or hypoesthesia of the anterior oral regions [10].

Effective coordination of the muscles involved in swallowing is essential in order to complete the oral and subsequent phases of swallowing, facilitating deglutition of the saliva that accumulates within the mouth. Inadequate swallowing habits with malpositioning of the head, together with local alterations in the form of defective lip sealing, give rise to salivary incontinence, which in turn worsens in the presence of excess saliva production.

Clinical Manifestations of Drooling

Drooling is a strong source of psychosocial stigmatization, limiting physical contact and sometimes leading to serious social isolation of the affected patient. Drooling can cause the wetting of clothing and bibs in the course of the day, with a need for continuous changing of clothes. In the severer cases, the floor and surrounding furniture can be affected as well [11].

The constant flow of saliva can keep the lips permanently humid, with irritation and excoriation of the skin around the mouth or chin. This situation in turn favors the development of fungal and bacterial infections in these areas – particularly in the form of candidiasis. In turn, basic functions such as speech and eating can be affected.

On the other hand, since the swallowing reflex is affected, saliva in some situations can be aspirated into the airway with respiratory tract infection such as pneumonia. In turn, there have been exceptional reports of dehydration secondary to excess water and electrolyte losses [1]. Likewise, the affected patients may present symptoms associated with their underlying systemic disease processes.

Evaluation of the severity of drooling presents difficulties, since there may be variations in salivary flow rate during the day and many influencing factors may be involved, such as medication, age, cognitive level and degree of attention. Observational and semiquantitative methods for assessing drooling have been described. Sochaniwskyj [12] developed a method based on the collection of leaked saliva using a chin cup during 30 min. This process was repeated 5 times to calculate an average and compensate for the variability of saliva leakage. The collected saliva was not whole saliva but only the saliva leaking from lips and reaching the chin [12].

Thomas-Stonell and Greenberg [13] in turn developed a classification of drooling with scores ranging from 1 to 5. Dry lips, with no drooling, corresponded to grade 1, while grade 2 was characterized by continuous lip wetting. Grade 3 in turn corresponded to wetting reaching the chin (moderate drooling), while grade 4 comprised the wetting of clothing around the neckline (severe drooling), and grade 5 corresponded to wetting of the clothes, hands and surrounding objects (profuse drooling) [13] (fig. 1, 2).

The drooling quotient is another semiquantitative method that has been validated for observ-

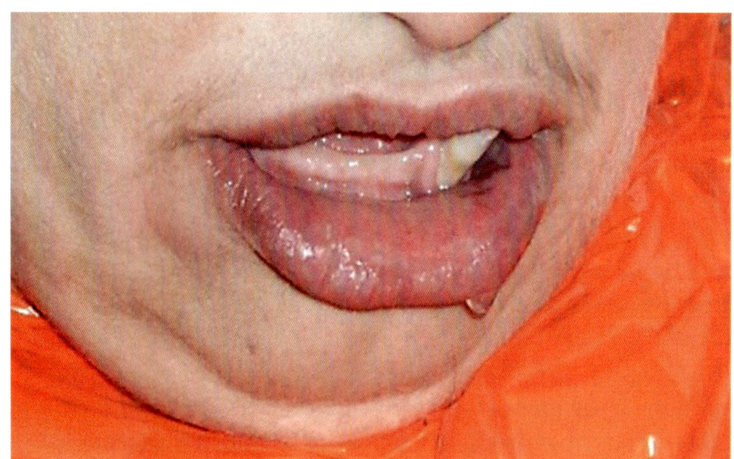

Fig. 1. Patient with clinical drooling type 2 [13].

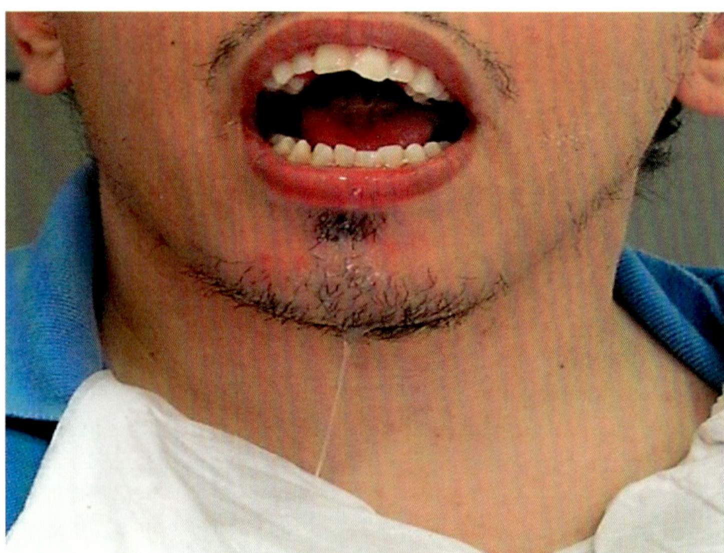

Fig. 2. Patient with clinical drooling type 4 [13].

ing the presence or absence of drooling on the lips and chin. It is assessed every 15 s during two 10-min periods separated by a 60-min break [14].

Management of Drooling

The clinical management of drooling is understood as the adoption of therapeutic strategies designed to correct or improve the patient neuromuscular imbalances and reduce the salivary flow rate. The problem is complex, however, because of the many intervening factors, and a multidisciplinary approach is needed, with the possible intervention of dentists, pediatricians, neurologists, surgeons, physiotherapists and speech therapists [15, 16].

A number of very distinct treatment options are available for dealing with drooling, from behavior-modifying techniques to orofacial myofunctional therapy, drug treatments, surgery and

Oral myofunctional therapy
    Lip-, chin- or throat-pressing techniques
    Stimulation plaque
    Feedback-based techniques
Behavior-modifying techniques
    Positive reinforcement
    Auditory feedback
    Humidity sensor control
Drug treatments
    Sublingual atropine
    Scopolamine transdermal patches
    Glycopyrrolate
    Benztropine
Botulinum toxin
Surgical methods
    Sectioning the parasympathetic neural pathway
    Submandibular gland tissue removal
    Submandibular duct translocation or repositioning

the use of botulinum toxin. Some treatments advocated in the past by a number of authors are no longer considered appropriate, such as radiotherapy for reducing salivary flow, due to the many side effects involved (table 3). Other options such as acupuncture or laser therapy have been little used and their results are moreover inconclusive; they consequently have not been included among the considered management options.

Oral Myofunctional Therapy
Oral motor therapy comprises a series of procedures and techniques that aim to improve oral function through the correction of bad habits. This type of therapy attempts to rehabilitate neuromuscular functions from an early age in order to favor adequate nasal breathing, correct mouth and lip sealing, and adequate neuromuscular control to improve chewing and food bolus formation and swallowing – thereby facilitating correct oral feeding and control of head positioning.

The intervention of the speech therapist may suffice to allow adequate saliva control. However, in many cases the lack of verbal and nonverbal understanding or the degree of mental retarda-

tion complicate communication and the application of such procedures.

Oral myofunctional therapies should be established on an individualized basis, using methods such as repetitive and coordinated exercises and lip-, chin- or throat-pressing techniques involving the stimulation of these anatomical points with vibrations and pressure. Acrylic plaque stimulation can also be used, according to the method developed by Castillo-Morales. These plaques contain certain elements which upon coming into contact with different areas induce neuromuscular stimulation. Examples of this include acrylic buttons applied to the palate to favor tongue retrusion, or acrylic striae at the lip level to stimulate lip sealing [17] (fig. 3).

Another feedback-based technique involves electromyographic monitoring of the muscle group which we wish to stimulate, placing two adhesive electrodes over the muscle for the recording of feedback. When the muscle contracts, the electromyographic registry informs of the change in activity by means of an acoustic or luminous signal. In this way the patient is made aware of the activity which he or she should carry out (e.g. swallowing), and certain elements of oral function can be corrected or improved, affording a positive impact upon training and improving oral motor function [18, 19].

Behavior-Modifying Techniques
Some clinicians consider that the first measure for controlling drooling is to improve patient postural tone – specifically referred to the trunk and head. This can be achieved by behavioral therapy. Positive reinforcement can be used to induce the patient to perform a given behavior such as swallowing, at certain moments. Combined use can be made of auditory feedback, conditioning the patient to swallow each time he or she hears a previously timed acoustic signal. We can also use a humidity sensor on the chin to produce a signal when the degree of humidity increases, instructing the patient to swallow.

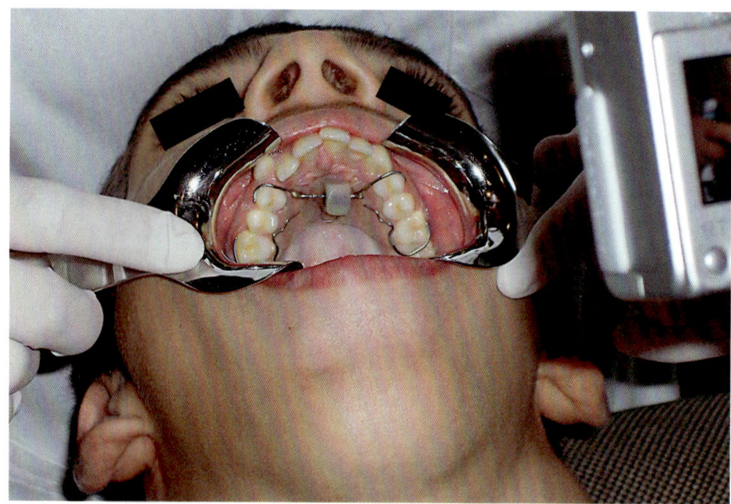

Fig. 3. Button applied to the palate for tongue retrusion [17].

These techniques have some inconveniences, including the considerable period of time needed to obtain results. Furthermore, the patients must have acceptable cognitive performance, with low or medium intensity drooling. In turn, these techniques must be repeated over time in order to reinforce the positive behaviors we wish to establish and maintain. In any case, the long-term results are uncertain, since the effects obtained decrease over time.

Drug Treatments

Pharmacological therapy indirectly controls drooling by reducing salivary secretion. The salivary glands are under the control of the autonomic nervous system, and salivation is mainly mediated by parasympathetic stimulation. Acetylcholine binds to the cholinergic muscarinic receptors to stimulate salivation, and a muscarinic receptor block induced by antagonist drugs gives rise to an important decrease in saliva output. For this reason anticholinergic agents have been used to control drooling. However, these drugs are contraindicated in many patients, such as those with narrow-angle glaucoma, heart problems, prostatic hypertrophy, paralytic ileus or other gastrointestinal disorders.

Sublingual atropine reduces drooling, but this treatment is contraindicated in patients with cognitive problems, as in dementia, and patients with psychotic symptoms (psychosis has been described due to atropine poisoning). Scopolamine is used in transdermal patches, causing important dry mouth, though it also produces important undesirable effects such as nausea, vomiting, diarrhea, irritability, mood changes and sleep disturbances. It has also been used as a nebulized formulation. These drugs are difficult to use because of their side effects.

Another drug substance that has been used is glycopyrrolate, which has a quaternary ammonium structure similar to that of atropine, with antimuscarinic effects [20]. It has been one of the most widely used drugs for the treatment of drooling. Glycopyrrolate has a long elimination half-life and does not cross the blood-brain barrier. However, it causes dry mouth, urinary retention, constipation and behavioral changes such as irritability. In turn, benztropine is a synthetic drug with antimuscarinic effects that can improve drooling control.

Botulinum Toxin

Botulinum neurotoxin type A has also been used for the control of drooling, particularly in children and young individuals with brain palsy. It is currently one of the most widely accepted treatment options, though approximately 10% of the patients fail to respond to this treatment [21–24].

Botulinum neurotoxin A inhibits the release of acetylcholine from the cholinergic nerve endings. It binds to proteins known as synaptosomal-associated protein 25, forming a complex that prevents exocytosis and inhibits fusion of the presynaptic vesicles containing the neurotransmitter. The studies published to date have reported uniformly positive results with the use of botulinum neurotoxin A.

The toxin is injected percutaneously into the parotid and submandibular glands under ultrasound guidance in order to improve the efficacy and safety of the technique. Two to three sites are usually injected in the parotid gland, together with 1–3 sites in the submandibular glands. The effect lasts for between 6 weeks and 6 months. This treatment can produce local side effects such as bruising, or even carotid artery or facial nerve damage. In turn, the action of the toxin upon the masticatory or pharyngeal muscles can cause chewing or swallowing problems. In children it can be administered under general anesthesia. Transductal injection is not advised, due to the risks involved.

Surgical Treatment
Surgery has been proposed for the control of drooling by many authors with the purpose of reducing salivary secretion. Numerous techniques have been proposed, including radical approaches such as gland tissue removal, salivary duct ligation or sectioning of the parasympathetic innervation of the gland. However, the most widely used techniques are those that cause fewer side effects, such as transpositioning of the salivary ducts.

These surgical techniques have caused many problems in the form of extreme dry mouth, loss of taste sensation, anterior tongue movement problems, gland swelling and a tendency towards severe sialoadenitis. All the possible alternatives must therefore be carefully evaluated before deciding to use such techniques.

Neurectomy implies sectioning the parasympathetic innervation at the chorda tympani level, to reduce salivary secretion of the submandibular and sublingual glands. The technique can be applied unilaterally or bilaterally, but there is a risk of causing hearing loss and diminished taste sensation in the anterior two thirds of the tongue. At present, neurectomy is considered too aggressive and does not solve the main cause of drooling [25].

The most widely used surgical techniques comprise translocation or repositioning of the glands, which are usually positioned at the arch or tonsillar fossa level – thereby guiding saliva posteriorly during the swallowing reflex [26, 27]. However, these procedures have also been associated with problems such as the appearance of ranulas or the loss of terminal sphincter smooth muscle function. Likewise, if saliva drainage is located in a very posterior position, caution is required due to the risk of aspiration.

Clinical Management of Drooling Patients
Drooling is a very common clinical problem in mentally disabled children and young individuals and to a lesser extent among elderly people with neurodegenerative diseases. In all cases, however, the condition is associated with an important loss of quality of life.

It is necessary to identify the factors underlying each individual case in order to define an adequate treatment strategy adapted to the needs of each concrete patient. In relation to the functional alterations in mentally disabled children, treatment should be started in the early stages of development, based on physiotherapy and myofunctional re-education, involving repetitive exercises, in an attempt to establish correct swallowing. In young individuals we must avoid and correct anterior open bite and other anterior malocclusions, attempting to stimulate correct lip sealing on closing the mouth. Likewise, botulinum toxin therapy can be used to lessen salivation.

When these treatments prove ineffective, or in adult patients, we can use antisialagogues such as atropinic agents, anticholinergic drugs and certain antihistamines. Surgical procedures should be avoided as far as possible, since in addition to their aggressive nature, they can only afford palliative effects.

Patients should be followed up regularly for encouraging the autonomous or assisted maintenance of correct oral hygiene, avoiding the appearance of infections, and treating them as soon as possible when they appear.

References

1 Meningaud JP, Pitak-Arnnop P, Chikhani L, Bertrand JC: Drooling of saliva: a review of the etiology and management options. Oral Surg Oral Med Oral Pathol Oral Radiol Endod 2006; 101:48–57.
2 Proulx M, de Courval FP, Wiseman MA, Panisset M: Salivary production in Parkinson's disease. Mov Disord 2005;20: 204–207.
3 Silvestre-Rangil J, Silvestre FJ, Puente-Sandoval A, Requeni-Bernal J, Simo-Ruiz JM: Clinical-therapeutic management of drooling: review and update. Med Oral Patol Oral Cir Bucal 2011; 16:e763–e766.
4 Felix DH, Luker J, Scully C: Oral medicine: dry mouth and disorders of salivation. Dental Update 2012;39:738–743.
5 Dodds WJ: Physiology of swallowing. Dysphagia 1989;3:171–178.
6 Tahmassebi JF, Curzon MEJ: Prevalence of drooling in children with cerebral palsy attending special schools. Dev Med Child Neurol 2003;45:613–617.
7 Chang SC, Lin CK, Tung LC, Chang NY: The association of drooling and health-related quality of life in children with cerebral palsy. Neuropsychiatr Dis Treat 2012;8:599–604.
8 Reid SM, McCutcheon J, Reddihough DS, Johnson H: Prevalence and predictors of drooling in 7- to 14-year-old children with cerebral palsy: a population study. Dev Med Child Neurol 2012;54: 1032–1036.
9 Blasco P: Prevalence and predictors of drooling. Dev Med Child Neurol 2012; 54:970.

10 Hedge AM, Pani SC: Drooling of saliva in children with cerebral palsy – etiology, prevalence, and relationship to salivary flow rate in an Indian population. Spec Care Dentist 2009;29:163–168.
11 Morales Chávez MC, Nualart Grollmus ZC, Silvestre-Donat FJ: Clinical prevalence of drooling in infant cerebral palsy. Med Oral Patol Oral Cir Bucal 2008; 13:E22–E26.
12 Sochaniwskyj AE: Drool quantification: noninvasive technique. Arch Phys Med Rehabil 1982;63:606–607.
13 Thomas-Stonell N, Greenberg J: Three treatment approaches and clinical factors in the reduction of drooling. Dysphagia 1988;3:73–78.
14 Reddihough D, Johnson H, Ferguson E: The role of a saliva control clinic in the management of drooling. J Paediatr Child Health 1992;28:395–397.
15 Daniel SJ: Multidisciplinary management of sialorrhea in children. Laryngoscope 2012;122:S67–S68.
16 Brei TJ: Management of drooling. Semin Pediatr Neurol 2003;10:265–270.
17 Limbrock GJ, Hoyer H, Scheying H: Drooling chewing and swallowing dysfunctions in children with cerebral palsy: treatment according to Castillo-Morales. ASDC J Dent Child 1990;57: 445–451.
18 Rodrigues dos Santos MT, Masiero D, Novo NF, Simionato MR: Oral conditions in children with cerebral palsy. J Dent Child (Chic) 2003;70:40–46.
19 Fairhurst CB, Cockerill H: Management of drooling in children. Arch Dis Child Educ Pract Ed 2011;96:25–30.

20 Mier RJ, Bachrach SJ, Lakin RC, Barker T, Childs J, Moran M: Treatment of sialorrhea with glycopyrrolate: a double-blind, dose-ranging study. Arch Pediatr Adolesc Med 2000;154:1214–1218.
21 Breheret R, Bizon A, Jeufroy C, Laccourreye L: Ultrasound-guided botulinum toxin injections for treatment of drooling. Eur Ann Otorhinolaryngol Head Neck Dis 2011;128:224–229.
22 Rodwell K, Edwards P, Ware R, Boyd R: Salivary gland botulinum toxin injections for drooling in children with cerebral palsy and neurodevelopmental disability: a systematic review. Dev Med Child Neurol 2012;54:977–987.
23 Jeung IS, Lee S, Kim HS, Yeo CK: Effect of botulinum toxin A injection into the salivary glands for sialorrhea in children with neurologic disorders. Ann Rehabil Med 2012;36:340–346.
24 Çiftçi T, Akinci D, Yuttutan N, Akhan O: US-guided botulinum toxin injection for excessive drooling in children. Diagn Interv Radiol 2013;19:56–60.
25 Mullins WM, Gross CW, Moore JM: Long term follow-up of tympanic neurectomy for sialorrhea. Laryngoscope 1979;89:1219–1223.
26 Squires N, Wills A, Rowson J: The management of drooling in adults with neurological conditions. Curr Opin Otolaryngol Head Neck Surg 2012;20:171–176.
27 Chanu NP, Sahni JK, Aneja S, Naglot S: Four-duct ligation in children with drooling. Am J Otolaryngol 2012;33: 604–607.

Francisco Javier Silvestre-Donat
Special Care Patients, Clinica odontológica universitaria
Gascó Oliag 1
ES–46010 Valencia (Spain)
E-Mail francisco.silvestre@uv.es

Ligtenberg AJM, Veerman ECI (eds): Saliva: Secretion and Functions.
Monogr Oral Sci. Basel, Karger, 2014, vol 24, pp 135–148 (DOI: 10.1159/000358794)

Salivary Gland Diseases: Infections, Sialolithiasis and Mucoceles

Konstantina Delli · Fred K.L. Spijkervet · Arjan Vissink

Department of Oral and Maxillofacial Surgery, University of Groningen, University Medical Center Groningen, Groningen, The Netherlands

Abstract

The three most frequently diagnosed salivary gland diseases are salivary gland infections, sialolithiasis and mucoceles. Salivary gland infections are usually of bacterial or viral etiology and can be divided into acute and chronic types. Occasionally they can result from obstruction of the salivary duct, an autoimmmune disease or cancer therapy. Infections can occur in all types of salivary glands and are observed at all ages. Sialolithiasis is characterized by the development of calcified structures in the salivary glands, especially in the submandibular gland. Sialoliths are generally attributed to retention of saliva and are usually accompanied by swelling and pain when a salivary stimulus is applied. Mucoceles can be differentiated into mucus extravasation phenomenon or mucus escape reaction, mucus retention cysts and ranulas. They result from extravasation of saliva into the surrounding soft tissues or from retention of saliva within the duct.

© 2014 S. Karger AG, Basel

Salivary gland diseases can be generally classified as neoplastic (benign and malignant) and nonneoplastic. The latter can be further subdivided into developmental, obstructive, inflammatory, infectious and autoimmune diseases (table 1). This paper will focus on the three most frequently observed salivary gland diseases, i.e. sialadenitis, sialolithiasis and mucoceles, with respect to their etiology, clinical manifestations and therapy.

Infections

Sialadenitis refers to acute, chronic or recurrent infections as well as inflammatory conditions affecting the salivary glands. This section will concentrate on analyzing the most common types of sialadenitis, i.e. acute bacterial, chronic bacterial and viral sialadenitis. Autoimmune sialadenitis (e.g., Sjögren's syndrome) and irradiation-induced sialadenitis (head and neck radiotherapy, treatment with radioactive iodine) are discussed in the paper by Delli et al. [this vol., pp. 109–125].

Acute Bacterial Sialadenitis
Acute bacterial sialadenitis (ABS) is caused by bacteria invading salivary glands, mostly as the result of an ascending infection originating from the oral cavity [1].

Table 1. Classification of salivary gland diseases

Neoplastic	Nonneoplastic
Benign	Developmental
Malignant	Obstructive
	Inflammatory
	Infectious
	Autoimmune

Saliva has antibacterial properties. Additionally, salivary flow removes debris and hinders bacteria to enter the duct. Furthermore, the general immune system and the physical characteristics of the duct and glandular system protect salivary glands from infections. In case a natural barrier is deranged, the risk of ABS is considerably increased.

ABS is most commonly observed in parotid glands. A plausible explanation is that the submandibular and sublingual salivary glands produce mucous saliva [2], which contains lysozyme, which in turn breaks the 1–4 link between N-acetylmuramic acid and N-acetylglucosamine in the bacterial cell wall. Besides, mucous saliva contains IgA, IgM, IgD and sialic acids, which inhibit bacterial attachment to epithelial cells.

Parotid Gland

Acute bacterial parotitis (ABP) can be divided into hospital-acquired and community-acquired types. *Hospital-acquired ABP* results from diminished salivary flow related to sepsis, acute illness or postoperative state, usually occurring between the 5th and 7th postoperative days [3]. In cultures often different flora can be found compared to community-acquired ABP. In hospital-acquired ABP, *Staphylococcus aureus* prevails in over 50% of cases, while intensive care unit patients are usually infected by hospital-acquired flora: *Einkenella corrodens, Escherichia coli, Fusobacterium* species, *Klebsiella* species, *Prevotella* species and/or *Pseudomonas* species. *Community-ac-*

quired ABP is more often diagnosed than the hospital-acquired type and is usually associated with *Staphylococcus epidermidis* and *Streptococcus* species [1, 4]. In the majority of cases community-acquired ABP is related to obstruction of the duct either due to a sialolith, mucous plug, adhesions, strictures or malformation of the excretory ducts. In cases with recurrent chronic parotitis with acute episodes, further diagnostics on the nature of the recurrence is mandatory (see next paragraph). Xerogenic medication, cheek biting, medical conditions (e.g. poorly controlled diabetes, hepatic disease and renal disease), malnutrition and dehydration can also promote the development of ABP.

Submandibular Salivary Gland

The most common cause of acute bacterial submandibular sialadenitis is obstruction of Wharton's duct by a sialolith (see section 'Sialolithiasis').

Minor Salivary Glands

Acute bacterial infection of minor salivary glands is a sparse phenomenon. A rare inflammatory disorder of the lips is cheilitis glandularis, which is characterized by swelling of the lip and hyperplasia of the labial salivary glands [5]. It usually affects the lower lip and shows a predilection for middle-aged and elderly men [6].

General Clinical Characteristics

The affected major salivary glands appear swollen and painful, while the overlying skin may be erythematous and warm (fig. 1). Some patients develop fever, and a purulent discharge is often detected from the duct when the gland is massaged (fig. 2). Regional lymph nodes are frequently enlarged.

Therapy

The patient is instructed to empty the salivary gland by massaging the affected parotid or submandibular gland from dorsal to ventral. To support the drainage of the affected gland, a salivary

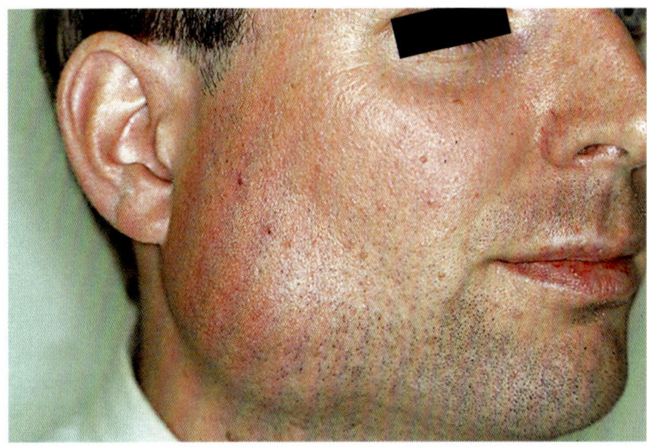

Fig. 1. ABS: swollen and painful parotid gland, with erythematous and warm overlying skin.

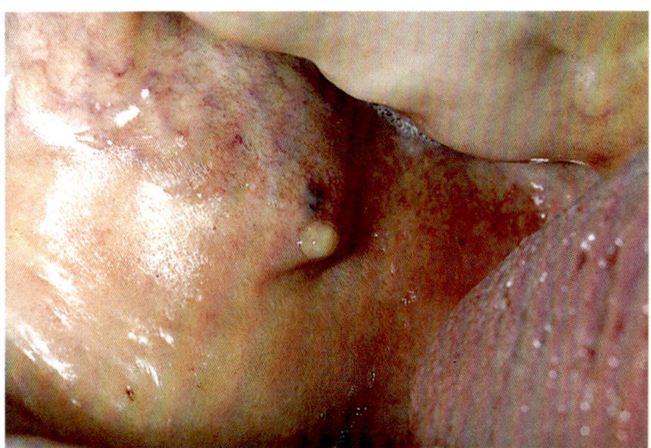

Fig. 2. ABS: purulent discharge detected from the duct when the gland is massaged.

stimulus can be applied, e.g. sucking on citric acid-containing lozenges. The goal of the therapy is to empty the gland and to restore salivary secretion as the bacterial infection is a retrograde infection originating from ascending bacteria from the orifice via the ductal system towards the secretory parenchyma.

When supportive antibiotic therapy is needed, i.e. in cases when massaging and stimulating the affected salivary gland is (partially) uneventful, antimicrobial coverage of staphylococcal and streptococcal species is necessary since most cases of ABS are caused by bacteria originating from the oral flora. Although it is recommended to commence antibiotic therapy after a culture from the affected duct, a β-lactam antibiotic is the initial drug of choice for most clinicians (amoxicillin 500 mg t.i.d.), whereas in hospital-acquired ABP amoxicillin/clavulanate (500/125 mg t.i.d.) is the drug of choice for most clinicians [3]. In case of allergy to penicillin, clindamycin (300 mg q.i.d.) is the generally preferred alternative. Recurrent bacterial infections and abscesses necessitate further investigation to rule out sialolithiasis fibrosis, and underlying systemic diseases.

Treatment of cheilitis glandularis involves sun protection, oral antibiotics in case of secondary infection, local corticosteroid injection, and vermilionectomy [7].

Chronic Bacterial Sialadenitis
Chronic bacterial sialadenitis (CBS) typically affects the parotid gland, mostly unilateral. CBS is usually attributed to ductal obstruction due to a mucous plug, stricture and/or adhesions, and in juvenile chronic parotitis additionally malformation of the excretory ducts or congenital duct ectasia should be considered. Also a history of recent pyogenic infection or mumps may be present. In the submandibular gland development of chronic sialadenitis is usually linked to the presence of a sialolith. A decrease in the saliva secretion rate is considered as the primary pathogenic event, which leads to stasis of secretions with resultant ascending bacterial invasion of the ductal system [8]. Maynard [9] supports that recurrent juvenile parotitis (see below) in about 30% of children eventually results in the development CBS.

CBS generally launches with an episode of ABS. The swelling may persist from days to months, and after resolving clinical quiescence may range from weeks to years. According to Carlson [1], two types of chronic bacterial parotitis are observed:

(i) the adult type, usually associated with *S. aureus*;

(ii) the juvenile type, linked to viridans streptococci.

Treatment of CBS should be focused on reducing the inflammation in the gland and on clearing the intraductal system from precipitated serum proteins. Short-term use of corticosteroids may be used to diminish inflammation. Salivary gland production can be enhanced by sialagogues whereas sialadenoscopy can facilitate both diagnosis and clearance of the duct, release of strictures or adhesions. Acute exacerbations should be treated as acute suppurative sialadenitis [8].

Adult Chronic Parotitis

Adult chronic parotitis (ACP) is a nonspecific sialadenitis characterized by unilateral or bilateral, and persistent or periodic swelling of the parotid glands [10]. Although the etiology of ACP is most likely multifactorial [11], it is supported that salivary stasis predisposes to inflammation which leads to multifocal wall irregularities, propagating into stricture formation [10]. ACP should not be mistaken for secondary parotitis due to Sjögren's syndrome, sialolithiasis, radiation treatment or treatment with radioactive iodine [1].

Clinically, ACP appears after an abrupt episode of parotid swelling, while usually no association with meals or seasons can be established, therefore excluding the possibility of obstruction or allergy.

Juvenile Recurrent Parotitis

Juvenile recurrent parotitis is a chronic condition involving recurrent inflammation of the parotid glands in children. Even though it is an infrequent disorder, it is the second most common salivary gland disease in children, after mumps [12]. Juvenile recurrent parotitis has a vague etiology: congenital abnormalities or strictures of Stenson's duct, history of viral or bacterial infection, allergy, local manifestation of an autoimmune disease, trauma or foreign body in the duct have often been associated [13].

Juvenile recurrent parotitis is characterized by usually unilateral recurrent episodes of swelling and/or pain along with fever and malaise. The

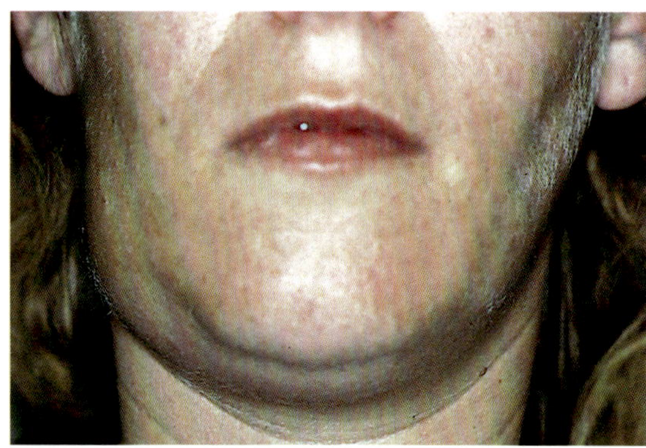

Fig. 3. Kuttner's tumor.

first episode typically occurs between the ages of 3 and 6 years [14]. An episode lasts often from 2 up to 12 weeks.

Treatment is as a rule symptomatic, and β-lactam antibiotics (e.g. penicillin V, amoxicillin-clavulanate) as well as analgesics are empirically used for 7 days to shorten the duration of the symptoms [15]. In selected cases long-term use of doxycycline for 8 weeks are advocated [10], also in combination with the use of corticosteroids.

Growing evidence advocates sialadenoscopy with high-pressure saline irrigation as the initial treatment of choice [13].

Chronic Sclerosing Submandibular Sialadenitis (Kuttner's Tumor)

Chronic sclerosing submandibular sialadenitis (CSSS) was first described by Kuttner in 1896 and represents an inflammatory process primarily involving the submandibular salivary glands. CSSS clinically appears as a painful swelling (fig. 3), while histologically it demonstrates a loss in acinar tissue, dense periductal fibrosis and lymphocytic infiltration with lymphoid follicle formation in some cases [16]. The most frequent etiological factor is considered to be sialoliths. Secretory disorders, infectious agents and immunological responses have been additionally associated with CSSS [17–20]. Abe et al. [21] demonstrated that patients with CSSS may potentially have an underlying systemic IgG4-related plasmacytic disease.

Viral Sialadenitis
Mumps
Mumps is by far the most common viral disease of the salivary glands, characterized by an acute bilateral nonsuppurative swelling of the parotid glands, although all salivary glands can be involved [22]. Mumps is caused by a single-stranded paramyxovirus with an affinity for the glandular epithelium. Viral replication in the parotid ductal epithelium (or the ductal epithelium of other salivary glands affected by this virus) leads to periductal interstitial edema and local inflammation with infiltrates of lymphocytes and macrophages [23]. Other viruses (e.g. coxsackievirus A and B, and Epstein-Barr virus, EBV) have also been linked to salivary gland disease.

Mumps is most frequently contracted during childhood, and more than 85% of cases occur in patients under the age of 15 [8]. After an incubation of 15–18 days, fever, malaise, headache and finally painful glandular swelling appear. Patients are considered contagious immediately before and/or at the beginning of the glandular swelling. Symptoms typically disappear spontaneously within 10 days, therefore treatment is only palliative [1].

Common complications of mumps include pancreatitis, meningoencephalitis, mastitis, orchitis, thyroiditis, myocarditis and hepatitis. The routine administration of measles-mumps-rubella vaccination, as advocated by the WHO and available in most countries worldwide, has decreased the incidence of mumps cases and is acknowledged as the most important advancement in the management of the disease.

Human Immunodeficiency Virus
The most prevalent salivary gland complication in human immunodeficiency virus (HIV)-infected patients is salivary gland swelling due to acute sialadenitis or HIV-associated salivary gland disease [24]. Salivary flow in HIV patients is reduced and salivary composition is changed, characterized by increased sodium, chloride, lysozyme, peroxidase, lactoferrin and IgA levels [25].

Parotid enlargement is estimated to occur in approximately 1–10% of HIV-infected patients [26] and develops usually secondary to benign lymphoepithelial cysts in the parotid gland. The most widespread term to describe this condition is 'diffuse infiltrative lymphocytosis syndrome (DILS)' despite the fact that several terms are used by clinicians and researchers. DILS is principally manifested as bilateral parotid painless swellings. The etiology of DILS is controversial and the most accepted theory supports that parotid enlargement results from proliferation of glandular epithelium that is trapped in the 5–10 embryologically derived lymph nodes in the gland [27, 28]. The phenomenon may be enhanced by ductal obstruction induced by lymphoid proliferation [29, 30]. DILS is extremely rare in the HIV-negative population; thus, the presence of these lesions necessitates investigation for HIV infection.

Ultrasound scan, computed tomography (CT) and magnetic resonance imaging (MRI) can assist in the diagnosis of DILS. The management of DILS is not very well established and usually involves close observation, repeated aspiration, antiretroviral medication, sclerosing therapy, radiation therapy and surgery.

Epstein-Barr Virus
EBV is a member of the Herpesviridae family responsible for infectious mononucleosis, nasopharyngeal carcinoma and Burkitt's lymphoma. EBV can also infect salivary glands. Saliva is the only bodily fluid known to contain viral particles of EBV. EBV replicates in the salivary glands and remains latent in affected individuals. It has been linked to the etiology of Sjögren's syndrome [31], because:

(i) it is present in salivary gland epithelial cells;

(ii) salivary gland biopsies of patients with Sjögren's syndrome contain increased EBV DNA compared to normal glands, indicating thus inability of lymphoid infiltration to control EBV replication;

(iii) salivary gland epithelial cells express high levels of HLA-DR antigens and EBV-associated antigens to immune T cells in patients with Sjögren's syndrome.

Nevertheless, a positive association has not (yet) been confirmed.

Due to the close correlation of EBV and malignancies (e.g. nasopharyngeal carcinoma, Burkitt's lymphoma, etc.) a possible involvement of EBV has been investigated in salivary gland tumors. Researchers report a close association between EBV and salivary gland malignancies [32], but the pathogenic mechanism remains to be elucidated.

Hepatitis C Virus

Hepatitis C virus (HCV) is a small enveloped, single-stranded RNA virus of the family Flaviviridae, causing hepatitis C in humans. Apart from hepatocytes and lymphocytes, HCV seems to show a tropism for lacrimal and salivary epithelial cells [33]. A chronic focal Sjögren's syndrome-like sialadenitis is observed in about 50% of HCV-infected patients [34], albeit sicca symptoms are less common and milder. Histologically, lymphocytic infiltrates are pericapillary (as opposed to the periductal infiltrate of Sjögren's syndrome), and lack of damage is typically observed [35]. Based on these facts, HCV has been regarded as one of the trigger viruses of Sjögren's syndrome. Interestingly some researchers demonstrated prevalence of HCV presence in primary Sjögren's syndrome patients [35], while others evidenced no difference between Sjögren's syndrome patients and healthy controls [36, 37]. However, according to the American-European classification criteria and the American College of Rheumatology classification criteria for Sjögren's syndrome, like for HIV and sarcoidosis, HCV infection is to date an exclusion criterion for diagnosing a patient having Sjögren's syndrome [38, 39]. One, however, always has to consider that Sjögren's syndrome might arise simultaneously to HCV, HIV or sarcoidosis [40].

Other Infections

Influenza and parainfluenza, coxsackie-, echo- and lymphocytic choriomeningitis viruses may be considered responsible for multiple episodes of mumps [8, 41]. The role of human papillomavirus and cytomegalovirus concerning the etiology of salivary gland tumors is under investigation. Sporadic reports entail the presence of human herpes simplex virus 8 in Warthin's tumors, but larger cohort studies are needed to determine this relationship [42].

Sialolithiasis

Sialoliths are calcified structures developing in salivary glands. Sialolithiasis is the second most common disease of the major salivary glands, after mumps. In countries applying routine administration of measles-mumps-rubella vaccination, the incidence of mumps has greatly decreased. Sialolithiasis accounts for 30% of all salivary disorders with an estimated incidence of 0.001–1.0% of the general population. Interestingly, sialolithiasis exhibits a higher incidence in males aged between 30 and 60 years [43]. Sialoliths are commonly located within the ductal system of the submandibular salivary gland (80–90%; fig. 4), followed to a lesser extent by the parotid gland (5–20%) [44]. The long upward path of the submandibular duct (Wharton's duct) as well as the thicker, mucoid secretions of this gland may be held responsible for the higher frequency of sialoliths in this site [45]. The sublingual and the minor salivary glands are rarely affected in humans.

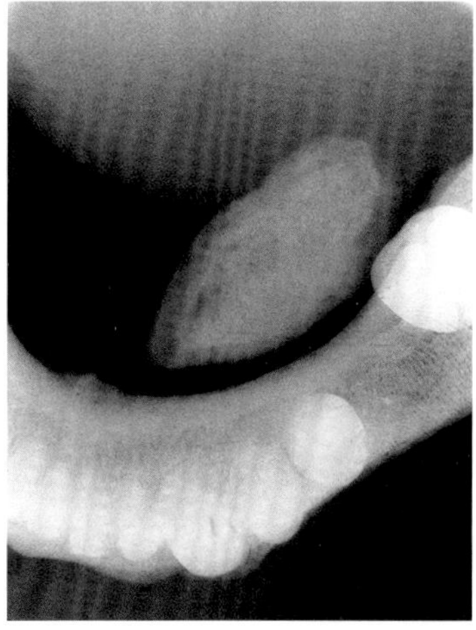

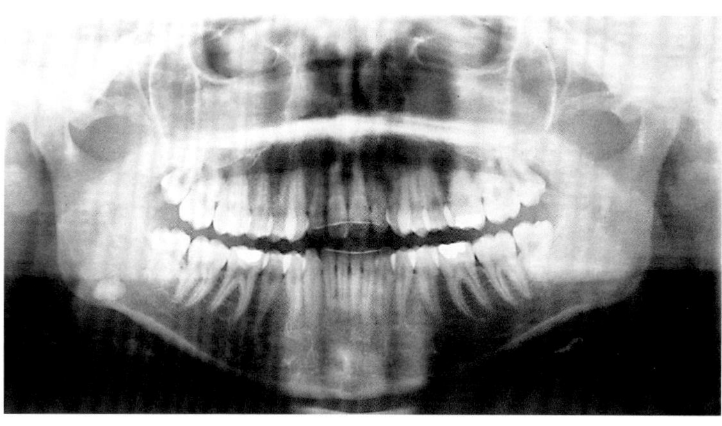

Fig. 4. Sialolithiasis: sialolith located within the ductal system of the submandibular salivary gland.

Fig. 5. Sialolithiasis: characteristic orthopantomogram with sialolith typically appearing as a radiopaque mass.

The composition of sialoliths may vary, but a mixture of amorphous carbonated calcium phosphate, carbonated apatite and whitlockite is often found. Apatites and/or proteins according to their history of creation and growth stratify the inner part of sialoliths. Mucins are abundantly identified and >60% of sialoliths contain these proteins. In fact, a noncalcified mucous plug may cause complaints resembling sialolithiasis. The spatial organization of sialoliths is complex. Some sialoliths have a purely proteic nucleus, surrounded by apatites, while others have a microcrystalline and lamellar body, with a varnished and fibrous crust. Irregular stratifications, in accordance with episodes of local infection and inflammation, are often observed [46].

Etiology
Sialolith formation can be generally attributed to retention of saliva due to morphoanatomical factors (salivary duct stenosis, salivary duct diverticuli, etc.) and to changes in salivary composition (high supersaturation, crystallization inhibitor deficit, etc.) [47].

Clinical Characteristics
Obstruction of salivary secretion by a sialolith can lead to swelling and pain and is usually associated with mealtimes since at that time salivary secretion is enhanced. The intensity of symptoms depends on the degree of obstruction and the amount of pressure exerted in the gland.

Diagnosis
A sialolith can be identified by radiographic examination, ultrasonography, sialography, CT, sialo-MRI and sialoendoscopy.

Radiographic Examination
On radiographic examination sialoliths typically appear as radiopaque masses (fig. 5). On panoramic radiographs they can appear superimposed on the mandible, while sialoliths located in the final part of the submandibular duct can be regularly recognized with an occlusal radiograph (fig. 4). Noncalcified ductal obstructions and sialoliths with a low degree of calcification may be invisible in regular radiographs.

Ultrasonography

Ultrasonography tends to replace radiographic examination in diagnosing sialolithiasis, since it can detect almost all intraparenchymal stones in the parotid and submandibular salivary gland. Nonetheless, ultrasonography sensibility decreases for ductal and papillary stones [21]. It is estimated that its sensitivity ranges from 59.1 to 93.7% whereas its specificity varies from 86.7 to 100% [48].

Sialography

Conventional sialography demonstrates sensitivity ranging from 64 to 100% and specificity from 88 to 100% [49]. Sialography, and especially in its digital modality, allows visualization of the main duct to the quaternary branches and the parenchyma of the gland. Furthermore, it can be supportive in the qualification of patients for sialoendoscopy and in selecting the strategy of the treatment procedure [50].

Computer Tomography

CT is a method of identifying sialoliths in cases of painful salivary glands and suspicion of few very small sialoliths [51]. This technique also enables for simultaneous imaging of the glandular tissue and salivary ducts. Nowadays in most situations cone beam CT is the preferred technique due to the outcome comparable to a conventional CT, but with a tenfold lower radiation dose.

Sialo-MRI

Sialo-MRI is an alternative technique, recognized as a highly effective, noninvasive diagnostic method of diagnosing sialoliths in patients with acute sialoadenitis. It identifies indirectly the presence of a stone by demonstrating proximal salivary duct dilation. Sialo-MRI allows simultaneous imaging of both the parotid and submandibular glands and assists in the differentiation of acute from chronic inflammation, based on the signal intensities [43].

Sialoendoscopy

In cases where all the aforementioned methods fail to provide a diagnosis, sialoendoscopy can be recruited for this objective (fig. 6). This method allows the clinician to have a direct visualization of the salivary gland lumen and if required to proceed at the same time as treatment. Both calcified and noncalcified stones as well as mucous plugs can be detected and removed.

Therapy

The treatment of sialolithiasis is principally dependent on the size of the sialolith.

Small sialoliths can be milked toward the duct by gentle massage. Sialagogues and increased consumption of acid-tasting fluids can contribute to this effort. In patients with acute symptoms, antibiotics and anti-inflammatory agents may well lead to a spontaneous stone expression through the papilla [52].

Larger sialoliths are usually removed surgically. Nevertheless, in cases of posteriorly located submandibular or parotid sialoliths, with a high risk of facial nerve paralysis, a conservative approach is usually adopted [53]. Under this spectrum, new techniques have been developed for removing relatively large sialoliths from the major salivary glands. The conventional treatment with open surgical or gland resection tends to be replaced by endoscopic and endoscopically assisted methods [54]. Extracorporeal shock wave lithotripsy has been shown to be very effective when applied in the appropriate cases [55].

Mucoceles

Mucus Extravasation Phenomenon – Mucus Escape Reaction

Mucoceles, exclusive of the irritation fibroma, are the most common benign soft tissue lesions of the oral mucosa. They result from rupture of a salivary gland duct and extravasation of mucin

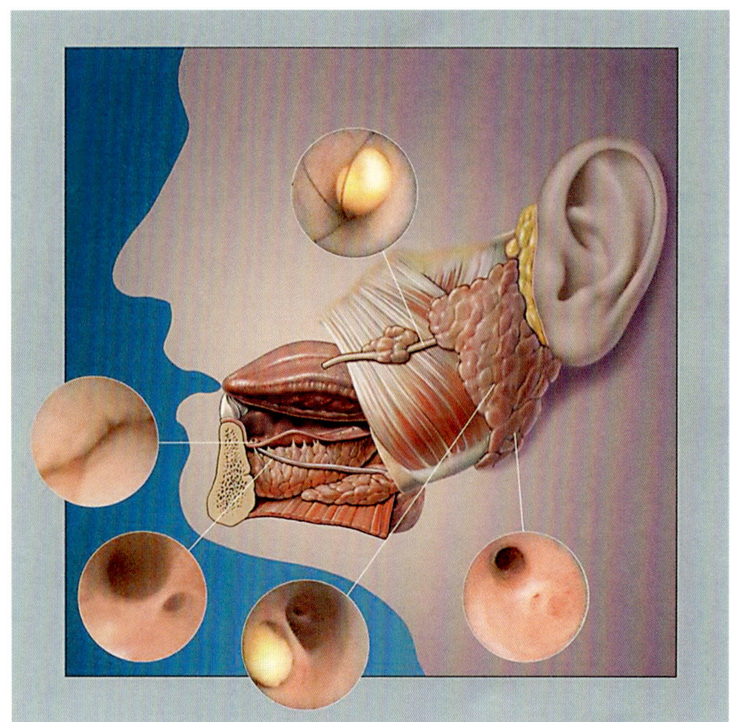

Fig. 6. Sialoendoscopic view of the ductal lumen of the parotid and submandibular salivary glands (courtesy of Prof. F. Marchal).

into the surrounding soft tissues. Although a history of local trauma is often reported, in some cases no obvious cause can be identified.

Clinical Features

Mucoceles of the extravasation type usually appear as painless spherical mucosal swellings ranging from 1–2 mm to several centimeters (fig. 7). Their color varies from deep blue to normal pink and depends on the size of the mucocele, the proximity to the surface and the elasticity of the surrounding tissues. They are typically fluctuant to palpation. Some patients report the periodic discharge of fluid from the lesion. Mucoceles are diagnosed in patients of all ages, but most cases arise in children and young adults [56]. The lower labial mucosa is definitely the most commonly involved anatomical location followed by the floor of the mouth and the buccal mucosa [45]. In striking contrast with the lower

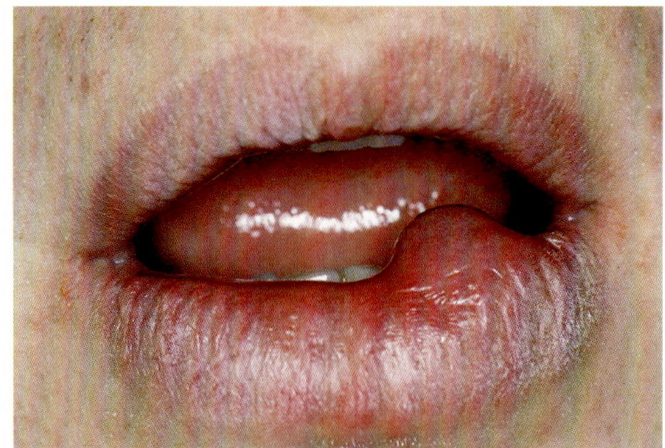

Fig. 7. Mucocele: spherical mucosal swellings with normal color located in the lower lip.

labial mucosa, mucoceles are extremely rare in the upper labial mucosa. In case of a swelling of the labial mucosa of the upper lip, the clinician first has to consider other etiologies including salivary gland tumors.

Histopathological Features
Mucoceles consist of spilled mucin surrounded by granulation tissue. The inflammation characteristically contains abundant foamy histiocytes, while it is not unusual that salivary gland tissue is recognized in the neighboring area. By definition, mucoceles do not exhibit a true cystic epithelial lining, but the condensed adjacent connective tissue can be misconceived for it.

Treatment
The prognosis of mucoceles is excellent. In some patients spontaneous rupture of the mucocele and automatic healing has been reported, but the vast majority of lesions has to be removed surgically. Surgical removal demands precise surgical skills, because of the close relation of the minor salivary glands to the sensible branches in the lower lip (fig. 8). Any projecting peripheral minor salivary glands should be excised in order to diminish the risk of recurrence. Incision of the mucoceles only leads to recurrence. Sutures should also be carefully placed in order to avoid injury of another adjacent gland. It is advised that the excised tissue should be submitted for microscopic examination to confirm the original diagnosis and to rule out the possibility of another salivary gland pathology.

Mucus Retention Cyst
Mucus retention cysts occur when a narrowed duct cannot adequately accommodate the exit of saliva, leading to ductal dilation and swelling.

Clinical Features
Mucus retention cysts are less frequent than mucoceles. They are reported to be more common in the elderly showing a slight predilection for women. In general they are located in the floor of the mouth, cheek and lower lip [57]. They can arise in both the major and minor salivary glands. Mucus retention cysts of the minor salivary glands may clinically look like mucoceles.

Sporadic reports of multiple mucus retention cysts exist and are possibly attributed to obstruct-

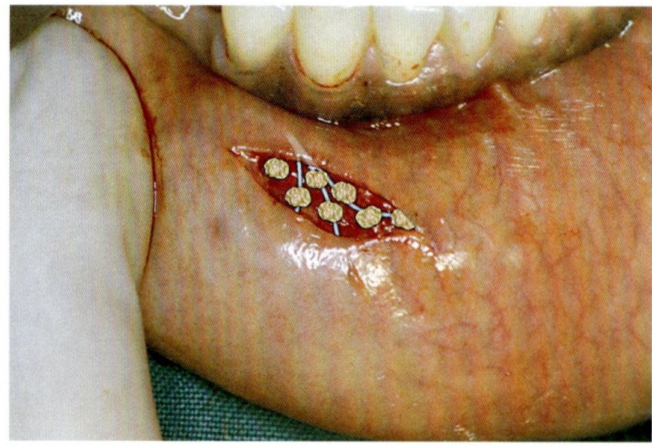

Fig. 8. The branch of the mental nerve that supplies the mucous membrane of the lip divides usually into two subbranches: a horizontal and a vertical. The latter has an ascending course toward the vermillion border and is in close relation to the labial salivary glands.

ed ducts by altered, persistent and viscid mucus or to congenital or acquired weakness in the structure of the ducts [58].

Histopathological Features
Unlike mucoceles, mucus retention cysts are true cysts lined by epithelium which varies from cuboidal to columnar or atrophic squamous. The epithelial lining surrounds thin or mucoid secretions. Ductal ectasia due to salivary duct obstruction is often observed and can be accompanied by oncocytic metaplasia of the epithelial lining. Eversole and Sabes [59] differentiate 3 histopathological types of mucus retention cysts:

(i) true mucus retention cyst, lined by nononcocytic ductal epithelium with minimal inflammation;

(ii) reactive oncocytoid cysts, exhibiting oncocytic metaplasia;

(iii) mucopapillary cysts, with a papillary multicystic growth pattern and regions of mucous metaplasia.

Treatment
Mucus retention cysts are mainly treated with surgical excision and have an excellent prognosis.

Ranula
Ranulas develop after extravasation of saliva from a ruptured main salivary duct or a ruptured acinus following obstruction. They are principally located in the floor of the mouth [56]. The source of the extravasation is usually the sublingual salivary gland. The term ranula is derived from the Latin word *rana*, meaning 'frog', since the lesion resembles a frog's belly. This term has also been unrepresentatively used for mucus retention cysts appearing after obstruction of the duct of the sublingual salivary gland or even for dermoid cysts and cystic hygromas [45, 60].

Clinical Features
Ranulas are more frequently seen in children and young adults. The lesion manifests as a gradually round or oval enlarging swelling of the floor of the mouth. In most cases it is fluctuant and nonmobile in palpation and usually situated lateral to the midline (fig. 9). Superficial lesions may be bluish to translucent in color, while deeper lesions may appear normal. Elevation of the tongue is observed in patients with ranulas reaching large dimensions.

Histopathological Features
The histopathological features of ranulas are similar to those of common mucoceles.

Treatment
Ranulas can be treated by: (i) marsupialization or (ii) surgical removal of the responsible gland.

The recurrence rates of the aforementioned treatment alternatives are estimated to be 66.7 and 1.55%, respectively [60].

Plunging Ranula
A plunging ranula is a ranula subtype exhibiting submandibular and/or submental swelling, with or without intraoral involvement (fig. 10). Several authors claim that plunging ranulas have a congenital etiology since they frequently appear in pediatric patients [61, 62], have a marked preponderance in Asians [63, 64] and in people with anomalies in the mylohyoid muscle [56, 65]. The differential diagnosis includes abscess, thyroglossal duct cyst, cystic hygroma, lymphangioma and lipoma.

Treatment
Treatment of plunging ranulas can be challenging and various different techniques have been described to date. According to Harrison [56] a successful treatment depends on:

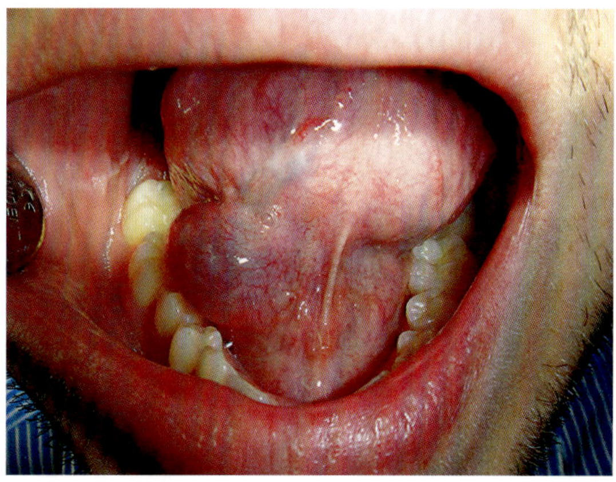

Fig. 9. Ranula: round bluish fluctuant enlarging swelling of the floor of the mouth, situated lateral to the midline.

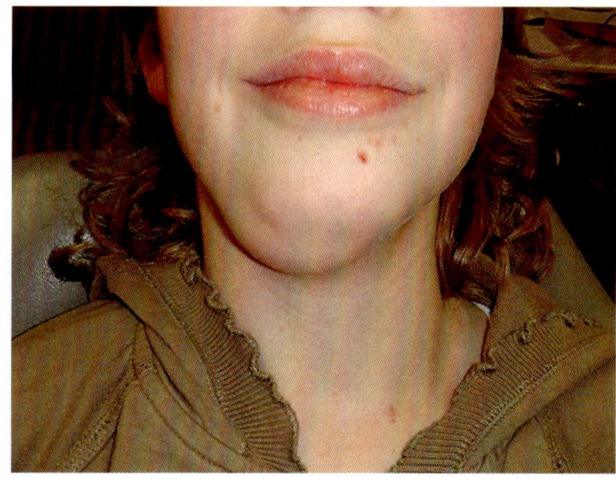

Fig. 10. Plunging ranula with extraoral involvement.

(i) removal of the part of the sublingual gland from which the extravasation occurs; this may involve either a complete excision or a conservative excision of only the responsible part of the gland;

(ii) induction of postoperative fibrosis to seal the leak or the inhibition of the secretory activity of the gland. In this way the balance between extravasation and the local opposition of macrophages and granulation and fibrous tissue changes in favor of the opposition.

Conclusion

Sialadenitis, sialolithiasis and mucoceles are 3 pathological entities frequently observed in the salivary glands. Their etiology can be obscure or multifactorial and the clinical features alarming and occasionally straddling. Thorough knowledge of these diseases and clinical experience are evenly important for the successful treatment of patients.

References

1 Carlson ER: Diagnosis and management of salivary gland infections. Oral Maxillofac Surg Clin North Am 2009;2:293–312.

2 Veerman EC, van den Keybus PA, Vissink A, Nieuw Amerongen AV: Human glandular salivas: their separate collection and analysis. Eur J Oral Sci 1996;104:346–352.

3 Katz J, Fisher D, Leviner E, Benoliel R, Sela MN: Bacterial colonization of the parotid duct in xerostomia. Int J Oral Maxillofac Surg 1990;19:7–9.

4 Nicolasora NP, Zacharek MA, Malani AN: Community-acquired methicillin-resistant *Staphylococcus aureus:* an emerging cause of acute bacterial parotitis. South Med J 2009;102:208–210.

5 Yanagawa T, Yamaguchi A, Harada H, Yamagata K, Ishibashi N, Noguchi M, Onizawa K, Bukawa H: Cheilitis glandularis: two case reports of Asian-Japanese men and literature review of Japanese cases. ISRN Dent 2011;2011:45756.

6 Leão JC, Ferreira AMC, Martins S, Jardim ML, Barrett AW, Scully C, Porter SR: Cheilitis glandularis: an unusual presentation in a patient with HIV infection. Oral Surg Oral Med Oral Pathol Oral Radiol Endod 2003;95:142–144.

7 Verma S: Cheilitis glandularis: a rare entity. Br J Dermatol 2003;148:362.

8 Bradley PJ: Microbiology and Management of Sialadenitis. Curr Infect Dis Rep 2002;4:217–224.

9 Maynard JD: Recurrent parotid enlargement. Br J Surg 1965;52:784–789.

10 Nahlieli O, Bar T, Shacham R, Eliav E, Hecht-Nakar L: Management of chronic recurrent parotitis: current therapy. J Oral Maxillofac Surg 2004;62:1150–1155.

11 Nichols RD: Surgical treatment of chronic suppurative parotitis. A critical review. Laryngoscope 1977;87:2066–2081.

12 Patel A, Karlis V: Diagnosis and management of pediatric salivary gland infections. Oral Maxillofac Surg Clin North Am 2009;21:345–352.

13 Galili D, Marmary Y: Juvenile recurrent parotitis: clinicoradiologic follow-up study and the beneficial effect of sialography. Oral Surg Oral Med Oral Pathol 1986;61:550–556.

14 Quenin S, Plouin-Gaudon I, Marchal F, Froehlich P, Disant F, Faure F: Juvenile recurrent parotitis: sialendoscopic approach. Arch Otolaryngol Head Neck Surg 2008;134:715–719.

15 Jain V, Mani NB, Singh M, Kumar L: Juvenile recurrent parotitis. Indian Pediatr 2000;37:1126–1129.

16 Seifert G, Miehlke A, Hubner K, Chilla R: Diseases of the salivary glands: Pathology, diagnosis, treatment, facial nerve surgery. Stuttgart: Georg Thieme Verlag 1986.

17 Harrison JD, Epivatianos A, Bhatia SN: Role of microliths in the aetiology of chronic submandibular sialadenitis: a clinicopathological investigation of 154 cases. Histopathology 1997;31:237–251.

18 Seifert G, Donath K: On the pathogenesis of the Kuttner tumor of the submandibular gland – analysis of 349 cases with chronic sialadenitis of the submandibular gland. HNO 1977;25:81–92.

19 Chan JK, Yip TT, Tsang WY, Poon YF, Wong CS, Ma VW: Specific association of Epstein-Barr virus with lymphoepithelial carcinoma among tumors and tumorlike lesions of the salivary gland. Arch Pathol Lab Med 1994;118:994–997.

20 Tiemann M, Teymoortash A, Schrader C, Werner JA, Parwaresch R, Seifert G, Klöppel G: Chronic sclerosing sialadenitis of the submandibular gland is mainly due to a T lymphocyte immune reaction. Mod Pathol 2002;15:845–852.

21 Abe T, Sato T, Tomaru Y, Sakata Y, Kokabu S, Hori N, Kobayashi A, Yoda T: Immunoglobulin G4-related sclerosing sialadenitis: report of two cases and review of the literature. Oral Surg Oral Med Oral Pathol Oral Radiol Endod 2009;108:544–550.

22 Schreiber A, Hershman G: Non-HIV viral infections of the salivary glands. Oral Maxillofac Surg Clin North Am 2009;21:331–338.

23 Weller TH, Craig JR: Isolation of mumps virus at autopsy. Am J Pathol 1949;25:1105–1115.

24 Shanti RM, Aziz SR: HIV-associated salivary gland disease. Oral Maxillofac Surg Clin North Am 2009;21:339–343.

25 Capaccio P, Monforte AD, Moroni M, Ottaviani F: Salivary stone lithotripsy in the HIV patient. Oral Surg Oral Med Oral Pathol Oral Radiol Endod 2002;93:525–527.

26 Dave SP, Pernas FG, Roy S: The benign lymphoepithelial cyst and a classification system for lymphocytic parotid gland enlargement in the pediatric HIV population. Laryngoscope 2007;117: 106–113.

27 Elliott JN, Oertel YC: Lymphoepithelial cysts of the salivary glands. Histologic and cytologic features. Am J Clin Pathol 1990;93:39–43.

28 Mandel L, Kim D, Uy C: Parotid gland swelling in HIV diffuse infiltrative CD8 lymphocytosis syndrome. Oral Surg Oral Med Oral Pathol Oral Radiol Endod 1998;85:565–568.

29 Ihrler S, Zietz C, Riederer A, Diebold J, Löhrs U: HIV-related parotid lymphoepithelial cysts. Immunohistochemistry and 3-D reconstruction of surgical and autopsy material with special reference to formal pathogenesis. Virchows Arch 1996;429:139–147.

30 Mandel L, Hong J: HIV-associated parotid lymphoepithelial cysts. J Am Dent Assoc 1999;130:528–532.

31 Fox RI, Luppi M, Pisa P, Kang HI: Potential role of Epstein-Barr virus in Sjögren's syndrome and rheumatoid arthritis. J Rheumatol Suppl 1992;32: 18–24.

32 Wen S, Mizugaki Y, Shinozaki F, Takada K: Epstein-Barr virus (EBV) infection in salivary gland tumors: lytic EBV infection in nonmalignant epithelial cells surrounded by EBV-positive T-lymphoma cells. Virology 1997;227:484–487.

33 Vitali C: Immunopathologic differences of Sjögren's syndrome versus sicca syndrome in HCV and HIV infection. Arthritis Res Ther 2011;13:233.

34 Loustaud-Ratti V, Riche A, Liozon E, Labrousse F, Soria P, Rogez S, Babany G, Delaire L, Denis F, Vidal E: Prevalence and characteristics of Sjögren's syndrome or sicca syndrome in chronic hepatitis C virus infection: a prospective study. J Rheumatol 2001;28:2245–2251.

35 Carrozzo M: Oral diseases associated with hepatitis C virus infection. 1. Sialadenitis and salivary glands lymphoma. Oral Dis 2008;14:123–130.

36 Vitali C, Sciuto M, Neri R, Greco F, Mavridis AK, Tzioufas A, Tsianos EV: Antihepatitis C virus antibodies in primary Sjögren's syndrome: false positive results are related to hyper-gamma-globulinaemia. Clin Exp Rheumatol 1992;10: 103–104.

37 Ramos-Casals M, De Vita S, Tzioufas AG: Hepatitis C virus, Sjögren's syndrome and B-cell lymphoma: linking infection, autoimmunity and cancer. Autoimmun Rev 2005;4:8–15.

38 Vitali C, Bombardieri S, Jonsson R, Moutsopoulos HM, Alexander EL, Carsons SE, Daniels TE, Fox PC, Fox RI, Kassan SS, Pillemer SR, Talal N, Weisman MH; European Study Group on Classification Criteria for Sjögren's Syndrome: Classification criteria for Sjögren's syndrome: a revised version of the European criteria proposed by the American-European Consensus Group. Ann Rheum Dis 2002;61:554–558.

39 Shiboski SC, Shiboski CH, Criswell L, Baer A, Challacombe S, Lanfranchi H, Schiødt M, Umehara H, Vivino F, Zhao Y, Dong Y, Greenspan D, Heidenreich AM, Helin P, Kirkham B, Kitagawa K, Larkin G, Li M, Lietman T, Lindegaard J, McNamara N, Sack K, Shirlaw P, Sugai S, Vollenweider C, Whitcher J, Wu A, Zhang S, Zhang W, Greenspan J, Daniels T; Sjögren's International Collaborative Clinical Alliance (SICCA) Research Groups: American College of Rheumatology classification criteria for Sjögren's syndrome: a data-driven, expert consensus approach in the Sjögren's International Collaborative Clinical Alliance cohort. Arthritis Care Res (Hoboken) 2012;64:475–487.

40 Van de Loosdrecht A, Kalk WWI, Bootsma H, Henselmans JM, Kraan J, Kallenberg CGM: Simultaneous presentation of sarcoidosis and Sjögren's syndrome. Rheumatology (Oxford) 2001;40:113–115.

41 Silvers AR, Som PM: Salivary glands. Radiol Clin North Am 1998;36:941–966.

42 Dalpa E, Gourvas V, Baritaki S, Miyakis S, Samaras V, Barbatis C, Sourvinos G, Spandidos DA: High prevalence of human herpes virus 8 (HHV-8) in patients with Warthin's tumors of the salivary gland. J Clin Virol 2008;42:182–185.

43 Andretta M, Tregnaghi A, Prosenikliev V, Staffieri A: Current opinions in sialolithiasis diagnosis and treatment. Acta Otorhinolaryngol Ital 2005;25:145–149.

44 Lustmann J, Regev E, Melamed Y: Sialolithiasis. A survey on 245 patients and a review of the literature. Int J Oral Maxillofac Surg 1990;19:135–138.

45 Neville BW, Douglas DD, Allen CM, Bouquot JE: Oral and Maxillofacial Pathology, ed 3. St Louis, Saunders, Elsevier, 2009.

46 Sabot JF, Gustin MP, Delahougue K, Faure F, Machon C, Hartmann DJ: Analytical investigation of salivary calculi, by mid-infrared spectroscopy. Analyst 2012;137:2095–2100.

47 Grases F, Santiago C, Simonet BM, Costa-Bauzá A: Sialolithiasis: mechanism of calculi formation and etiologic factors. Clin Chim Acta 2003;334:131–136.

48 Jäger L, Menauer F, Holzknecht N, Scholz V, Grevers G, Reiser M: Sialolithiasis: MR sialography of the submandibular duct – an alternative to conventional sialography and US? Radiology 2000;216:665–671.

49 Rzymska-Grala I, Stopa Z, Grala B, Gołębiowski M, Wanyura H, Zuchowska A, Sawicka M, Zmorzyński M: Salivary gland calculi – contemporary methods of imaging. Pol J Radiol 2010; 75:25–37.

50 Hasson O: Sialoendoscopy and sialography: strategies for assessment and treatment of salivary gland obstructions. J Oral Maxillofac Surg 2007;65: 300–304.

51 Madani G, Beale T: Inflammatory conditions of the salivary glands. Semin Ultrasound CT MRI 2006;27:440–451.

52 Marchal F, Dulguerov P: Sialolithiasis management: the state of the art. Arch Otolaryngol Head Neck Surg 2003;129: 951–956.

53 Dulguerov P, Marchal F, Lehmann W: Post-parotidectomy facial nerve paralysis: possible etiologic factors and results with routine facial nerve monitoring. Laryngoscope 1999;109:754–762.

54 Witt RL, Iro H, Koch M, McGurk M, Nahlieli O, Zenk J: Minimally invasive options for salivary calculi. Laryngoscope 2012;122:1306–1311.

55 Escudier MP, Brown JE, Putcha V, Capaccio P, McGurk M: Factors influencing the outcome of extracorporeal shock wave lithotripsy in the management of salivary calculi. Laryngoscope 2010;120: 1545–1549.

56 Harrison JD: Modern management and pathophysiology of ranula: literature review. Head Neck 2010;32:1310–1320.

57 Chi AC, Lambert PR 3rd, Richardson MS, Neville BW: Oral mucoceles: a clinicopathologic review of 1,824 cases, including unusual variants. J Oral Maxillofac Surg 2011;69:1086–1093.

58 Tal H, Altini M, Lemmer J: Multiple mucous retention cysts of the oral mucosa. Oral Surg Oral Med Oral Pathol 1984;58:692–695.

59 Eversole LR, Sabes WR: Minor salivary gland duct changes due to obstruction. Arch Otolaryngol 1971;94:19–24.

60 Zhao YF, Jia Y, Chen XM, Zhang WF: Clinical review of 580 ranulas. Oral Surg Oral Med Oral Pathol Oral Radiol Endod 2004;98:281–287.

61 Matt BH, Crockett DM: Plunging ranula in an infant. Otolaryngol Head Neck Surg 1988;99:330–333.

62 Clyburn VL 3rd, Smith JE, Rumboldt T, Matheus MG, Day TA: Ascending and plunging ranulas in a pediatric patient. Otolaryngol Head Neck Surg 2009;140: 948–949.

63 Davison MJ, Morton RP, McIvor NP: Plunging ranula: clinical observations. Head Neck 1998;20:636–638.

64 Morton RP, Ahmad Z, Jain P: Plunging ranula: congenital or acquired? Otolaryngol Head Neck Surg 2010;142:104– 107.

65 Hopp E, Mortensen B, Kolbenstvedt A: Mylohyoid herniation of the sublingual gland diagnosed by magnetic resonance imaging. Dentomaxillofac Radiol 2004; 33:351–353.

Konstantina Delli
Department of Oral and Maxillofacial Surgery
PO Box 30001
NL–9700 RB Groningen (The Netherlands)
E-Mail k.delli@umcg.nl

Author Index

Subject Index